William Holme Van Buren, Charles Morel

Compendium of Human Histology

William Holme Van Buren, Charles Morel

Compendium of Human Histology

ISBN/EAN: 9783337365417

Printed in Europe, USA, Canada, Australia, Japan

Cover: Foto ©berggeist007 / pixelio.de

More available books at **www.hansebooks.com**

COMPENDIUM

OF

HUMAN HISTOLOGY,

BY C. MOREL,

PROFESSOR AGRÉGÉ À LA FACULTÉ DE MEDECINE DE STRASBOURG.

Illustrated by Twenty-eight Plates.

TRANSLATED AND EDITED BY

W. H. VAN BUREN, M.D.,

PROFESSOR OF GENERAL AND DESCRIPTIVE ANATOMY IN THE UNIVERSITY OF NEW YORK;
MEMBER OF THE PATHOLOGICAL SOCIETY OF NEW YORK, &C., &C.

NEW YORK:

BAILLIÈRE BROTHERS, 440 BROADWAY.

LONDON:
H. BAILLIERE,
219 REGENT ST.

PARIS:
J. B. BAILLIERE ET FILS,
RUE HAUTEFEUILLE.

MELBOURNE:
F. BAILLIERE.

MADRID:
C. BAILLY-BAILLIERE,
CALLE DEL PRINCIPE.

1861.

R. CRAIGHEAD,
Printer, Stereotyper, and Electrotyper,
Caxton Building,
81, 83, and 85 *Centre Street.*

EDITOR'S PREFACE.

I HAVE prepared M. Morel's Compendium of Histology for the use of the American medical student in consequence of the excellence and fidelity of its plates, and the clear and concise manner in which all that is positively known of the science, up to the present moment, is set forth in the text.

An original work of the same character and similarly illustrated would involve a much greater expense, even if the same degree of merit could have been attained.

Histology, at the present day, is the progressive department of anatomical science, and its rapidly accumulating facts form the basis of modern physiology and pathology. Still in its youth, it is advancing steadily from year to year, and to keep pace with its progress, we must draw constantly upon all reliable sources of information. In our young and busy country the laborers are as yet too few, and too much, of necessity, employed in the active and practical duties of the medical profession, to institute original and systematic researches in minute anatomy to any great extent, and hence we are mainly indebted for our knowledge of it to our brethren of the old world.

To solicit the attention of the student to the subject of general anatomy, and to furnish him with an attractive text-book in a less elaborate form than the excellent works of Todd and Bowman, the Cyclopædia of Anatomy, &c., &c., is the object with which I have prepared the present volume for the press. I trust it may serve the purpose of an introduction to the more extensive works on the subject, and also to the numerous and faithful laborers in this department of science in Europe, and especially in Germany, where it is most successfully cultivated.

NEW YORK, *October*, 1860.

CONTENTS.

CHAPTER VI.

CHAPTER VII.

CHAPTER VIII.

CHAPTER IX.

CHAPTER X.

HUMAN HISTOLOGY.

INTRODUCTION.

THE science of Histology has for its object the study of the organic elements of the human body, in reference to their forms, the various changes which they undergo, and the part which they perform in the construction of its several tissues and organs.

In the present state of the science all of the simple elements of which the body is composed may be reduced to one of the following typical forms, viz: 1st, structureless material; 2d, cells; 3d, fibres; 4th, crystalline substance.

Structureless material, or amorphous substance, exists either as a liquid or solid; in the former condition it is of constant occurrence, in the latter it constitutes the basis or fundamental substance of several of the tissues, e.g. bone, cartilage, etc.

The cell, in the largest acceptation of the term, is a vesicle, varying greatly in shape, as well as in size, and consisting essentially of an envelope, and contained material of diverse appearance and nature.

1

The fibre, also variable in size, is either homogeneous throughout (connective fibre), or it takes the form of a tube, the walls of which are readily distinguishable from its contents (muscular fibre, nerve fibre, etc.).

In the normal state of the human body crystalline substance has, up to the present time, been found only in the internal ear, where it constitutes the otoliths, and in the encephalon; it is not positively determined that the concretions, composed of concentric layers, which are found in the pineal gland, are crystalline in their nature.

CHAPTER I.

Cells and Epithelial Membranes.

SECT. I. The simple cell is the organ which above all others is especially endowed with vital power, it is the formative element, in fact, of all the simple tissues, and is therefore, of necessity, our first object of study.

In every perfect cell we recognise the following parts: (1) A containing membrane, transparent, structureless, and exceedingly thin and delicate—the *cell-wall;* within this envelope (2) a liquid substance, generally transparent and granular, surrounding another vesicle, which usually presents a more strongly marked outline and thicker walls than those of the cell, called the *nucleus*, or *cyto blast;* and finally, (3) in the midst of the granular contents of this latter, we can detect, ordinarily, a granular body larger than the rest, known as the *nucleolus*. (Pl. I. fig. II. 1, 2, 3. Pl. II. fig. VI, VII. Pl. XIII, fig. I).

Structure of the cell.

Whenever these constituent parts cannot be recognised in a cell, it is to be inferred that it has already undergone transformation from its original condition, as we find to be the case, for example, in blood globules (Pl. I. fig. I. 1), and fat cells (Pl. I. fig. V.)

The contents of the cell, exclusive of its *nucleus* and *nucleolus*, have been described as ordinarily liquid, transparent, and finely granular. Sometimes, however, these transparent granules are replaced, entirely or in part, by minute masses, or grains, of an opaque and very dark colored substance (pigment), such as are to be seen, for example, in the pigment cells of the choroid, of the iris, and in many nerve-cells. (Pl. II. fig. I., II., Pl. XIII. fig. I. 6.)   There are cells, also, which in their normal condition contain numerous minute spherical globules with a clearly defined, dark outline, and possessing a very highly refractive power; these little pearl-like bodies are nothing more than free fat, and are known as *oil-globules*. Hepatic cells always contain a variable quantity of them, and globules of *colostrum* are filled with them.   (Pl. I. fig. III., Pl. XVIII. fig. VI. 2.) The same is true of the cells of sebaceous glands.

The presence of oil globules in a cell which normally contains none, is evidence of its approaching degeneration, and indicates arrest, or, at least, temporary perversion of its physiological development. This is to be seen in the pulmonary epithelial cells while tubercular deposit is taking place; it is also the anatomical lesion of the cells of renal epithelium in Bright's disease. Finally, crystals are sometimes formed in the cells of adipose tissue.   (Pl. II. fig. IV.)

When cells, and especially young cells, are subjected to the action of acetic acid under the microscope, both the cell wall and its contents very soon begin to grow pale, but their nuclei become more distinct; after a time the cell wall melts down and disappears, but the nucleus retains its natural appearance.   If, in place of acetic acid, caustic potash be applied in the same manner, even though largely diluted, the cell begins at once to swell up, grows pale, and finally disintegrates entirely; its elementary molecules, or granules, alone seem to possess the property of resisting the action of this chemical agent.

Although varying greatly in shape, all cells may be arranged under one of the following distinctive types: 1st, spherical cells; 2d, many-sided cells; 3d, scales; 4th, conical or cylindrical cells; 5th, ciliated cells; 6th, fusiform cells; 7th, branching, or star-shaped cells. *Different forms of cells.*

To the first class belong the embryonic ovule, and the cells developed directly from it; newly formed cells in the adult, and, in general, all cells which float in a liquid medium. *Globular or spherical cells.*

The second class includes the deeper seated cells of many of the epithelial membranes; the epithelial cells of glands composed of clustered follicles, and of some glands of tubular structure. Cells in the form of scales are found only in the superficial layer of the epidermis and epithelium of the tongue. *Polygonal or many-sided cells and scales.*

The deep layer of almost all the epithelial membranes which present a stratified arrangement of their cells, the epithelium of the intestinal canal, and of its tubular follicles, and that of a large proportion of the excretory ducts of glands, are made up of conical or cylindrical cells, (Pl. XXIII. fig. II. 3, Pl. XXVI. fig. VI. fig. XI.) *Conical cells.*

The free surfaces of the epithelial cells lining the walls of the air passages, uterus, fallopian tubes, etc., are furnished with a great number of minute hair-like projections (called *cilia* from their resemblance in shape to an eye-lash) which possess during life a peculiar vibratile motion, always in the same direc- *Ciliated cells.*

tion; these constitute the fifth class, and are known as ciliated cells.   (Pl. I. fig. VII.)

Fusiform cells.   Fusiform cells are found principally in tissues of recent origin which are undergoing fibrous transformation; cicatrices are formed by this class of cells, and some tumors are composed almost entirely of them.   (Pl. IV. fig. IV.)

Star-shaped or branching cells.   The last class comprises those cells in which the cell wall is developed into tubular or filiform prolongations or branches.   Examples: Most nerve cells of the nervous centres and ganglia; the cells of the external surface of the choroid; bone cells, plasmatic cells, etc.   (Pl. XIII. fig. I., fig. II., Pl. V. fig. IV., Pl. III. fig. IV.)

Every cell must derive its origin from another previously existing cell.   In the present state of science but two modes are known in which cell generation is accomplished in human histology: *endogenous* generation, and multiplication by *cleavage*.

In *endogenous* generation the process is not always exactly the same; sometimes the nucleus of the primitive cell developes itself into two secondary nuclei, each of which on the disappearance of their common envelope, surrounds itself by a portion of the granular contents of the cell, and a new cell wall making its appearance on its surface, (Kölliker), the infant cell is thus completed; (segmentation of the yelk).   In other cases the young nuclei, instead of making their appearance first in the interior of the nucleus of the primitive cell, develope themselves directly from the granular contents, and then grow into perfect cells in the mode already described; e.g. cells of fœtal marrow.   (Pl. VI. fig. IV.)

The mechanism of the process of generation by *cleavage* is as follows:

The primitive nucleus developes itself into two secondary nuclei, or, two nuclei may exist, from* its formation, in the primitive cell; then the cell wall contracts like an hour-glass enclosing a nucleus in either end, and finally a separation takes place at the contracted portion, the result of this metamorphosis being two perfect new cells; (Ex. cells of cartilage, cells of epithelium of intestine).

SECT. II. EPITHELIAL MEMBRANES.—The term Epithelium is applied to a class of membranes formed exclusively of cells, and ordinarily very thin and delicate. They invest all of the free surfaces which the body presents; thus the external integument is everywhere clothed with an expansion of epithelium, or epidermis, and the same is true of all mucous, serous, and synovial membranes, and of the membranes lining the cavities of the secreting glands, blood-vessels, and lymphatics. In view of the different shapes of the cells of which epithelial membranes are composed, they are divided into three groups, or classes: 1st, polygonal or scaly epithelium; 2d, cylindrical or conical; and 3d, ciliated epithelium.

Epithelial cells are in some instances spread out so as to form a simple lamina; in others they are found in several layers, one superimposed upon another. The first mode of arrangement constitutes *simple epithelium;* the second is known as *stratified epithelium.*

Heretofore the study of cells and epithelial membranes has been too much neglected; and yet there are, in truth, no histological elements of more importance—from whatever point of view they may be regarded. For is not the simple cell the constituent or formative element of all the normal tissues, as well as of those which result from diseased action? And of the latter, those over which our remedies exert the least power are composed almost exclusively of cells. Still farther: those organs of the body which hold the highest rank in view of the importance of their functions, which, in other words, possess the greatest amount of vital force, consist in the greatest proportion of cells; whilst the elementary fibre, and the organs mainly formed by it, perform functions which are simply mechanical.

CHAPTER II.

Fibres; Connecting Tissue.

The essential elements of connecting tissue* are fibres, and cells. Its fibres are of two kinds, viz: connective fibres properly so called, and elastic fibres. Its cells are diminutive in size, generally branched, but sometimes fusiform, and have received from Virchow† the name of *plasmatic* cells.

The elementary connective fibre is so exceedingly delicate in its proportions that it is impossible to measure its dimensions. Generally collected in fasciculi or bundles, they run parallel with each other, their outlines showing a slightly wavy or undulating disposition. In certain organs, tendons, for example, all the fasciculi of connective fibres are parallel to each other. (Pl. III. fig. I.; fig. III. 1, Pl. IV. fig. I. 1; fig. II. 1.) In the aponeuroses, the skin, the mucous, serous, and synovial membranes, they interlace so as to form a tissue or web, with meshes of varying size. (Pl. II. fig. IX.) These facts are readily demonstrated by examining a small fragment or slice from the surface of an aponeurosis or tendon, care

Connective fibres.

---

* The term *connecting tissue* is employed throughout the present work to designate that tissue heretofore generally known as *cellular, areolar* or *filamentous* tissue.—(*Ed.*)

† Rodolphe Virchow, Professor of Pathological Anatomy in the University of Berlin.—(*Ed.*)

being taken to cut in the direction of the fibres of the latter.

Elastic fibres are of larger size than those just described; the smallest, measure $\frac{1}{1500}$th of a line in breadth, but they may reach $\frac{1}{165}$th of a line (elastic coat of veins, Pl. XVI. fig. IV. 1). Their outlines are clearly marked by one, and more frequently by two black lines, between which is to be seen an entirely unorganized and transparent substance. They give off branches, also, in every direction, and these divisions uniting again with each other, form a network of variable closeness. Ordinarily, the principal branches of a fasciculus of elastic fibres run parallel with each other, as in the yellow elastic ligaments of the spine, (Pl. III. fig. V. 1); but the secondary branches which they give off present very well marked undulations, and most frequently curl upon themselves. (Pl. III. fig. V. 2; Pl. XV. fig. VI. 1.) Following this description it is hardly possible to confound elastic and true connective fibres with each other, but we possess additional means of bringing out their distinctive characteristics, by the use of certain chemical reagents. Thus, when we subject the connective fibre to the action of acetic acid, it becomes so pale as to be unrecognisable, and finally dissolves entirely in the liquid; the same result follows, and even more rapidly, on the application of diluted caustic potash. These reagents produce no alteration whatever in the appearance of the elastic fibres; dilute caustic potash is even employed in preparing clean and perfect specimens of them. To effect this, a piece of yellow elastic ligament is to be boiled for fifteen or twenty minutes in water containing some of the alkali.   All of the other elements which enter into the composition of the ligamentous tissue are dissolved, whilst the elastic fibres remain unchanged.

The cellular element of connecting tissue (the plasmatic cell) is a recent discovery; we are indebted to Virchow for the first thorough exposition of its nature, and especially of its important pathological relations.*

Plasmatic cells are minute corpuscles, sometimes fusiform, but more frequently star-shaped, with sharp outlines, and connected with each other by means of their branching prolongations, so as to constitute a network similar to that formed by the cells of bone. (Pl. III. fig. II. 1; fig. III. 2; fig. IV. 3.)   *Plasmatic cells.*

In tendons, we find plasmatic cells arranged in a longitudinal series, between the fasciculi of connective fibres.   (Pl. III. fig. II. and III.; Pl. IV. fig. I.)   In the skin and mucous membranes they are distributed

---

* It will have been already noticed by the reader, familiar with the present attitude of Histology in Germany, that our author has fully adopted the somewhat novel views of the celebrated Professor of Berlin. The assertion in the preceding chapter, that every cell must take its origin from a previously existing cell, is one of the new doctrines of Virchow, who denies that cells ever make their appearance by spontaneous generation, according to the generally received pathology of the present day, in an appropriate blastema or exudation, and insists that they are always produced by endogenous growth, or by fissure and cleavage of preëxisting nuclei and cells.

The discovery of the so-called plasmatic cell, as an element of connecting tissue, was first announced by Virchow at Wurzburg, where he was then professor, in 1851, and these cells play a most important part in the new pathological views which have since been so ably developed in his more recent and elaborate work on "Cellular Pathology," published at Berlin in 1858. Their existence was at first disputed by Henle, but admitted by Kölliker, Leydig of Wurzburg, Weber of Bonn, and most other German authorities.—(*Ed.*)

more irregularly. In studying the cells in tendon, both longitudinal and transverse sections must be examined, in order to see them well; their branching disposition can be clearly recognised only in the latter. (Pl. III. fig. IV. 3.)

Plasmatic cells are most advantageously studied, however, in the cornea; its entire substance, between the two layers of epithelium which invest its anterior and posterior surfaces, consists of an amorphous material in which we find myriads of star-shaped cells arranged in regular concentric lines running parallel with its surfaces. The numberless branches given off by the cells, on every side, anastomose in such a manner as to form a very beautiful network. Very dilute acetic acid must be applied to the section under examination; if the acid is too strong the branches of the cells are rendered invisible, and the cells themselves appear simply fusiform, or spindle-shaped. (Pl. II. fig. V.)

In the way of normal development, these plasmatic cells may transform themselves into cartilage cells, as in some tendons in old age, the inferior extremity of the *tendo achillis*, for example, and the cartilaginous enlargement of the *peronæus longus*, where it plays over the cuboid bone. This transformation is effected by the disappearance of the branches of the cell, and the production of an external envelope, which thickens into cartilage. (Pl. IV. fig. I. 2; fig. II. 2.)

They are metamorphosed in a like manner, in the periosteum, into bone cells, by investing themselves with a coating of earthy salts; it is also through the agency of these cells that layers of new bone are deposited by the periosteum in certain stages of periostitis. Senile opacity of the cornea is likewise explained by the appearance of oil-globules in the interior of its plasmatic cells. Finally, the researches of Virchow tend to prove that all, or nearly all, of the morbid formations developed in the meshes of the connecting tissue throughout the body, are traceable to the perverted growth of plasmatic cells.

Connecting tissue, then, is made up of the three elements just studied, mingled together in variable proportions, and, with them, of vessels and nerves, also variable in number. But it is to be observed that these latter exist in connecting tissue as accessory elements only. Thus, the numerous vessels habitually formed in certain layers of connecting tissue, as for example in that which underlies the mucous membrane of the intestinal canal, or some portions of the external integument, have nothing whatever to do with the nutrition of the connecting tissue around them; they merely pass through it to their ultimate destination, and are proportionate in number to the importance of the function to which they minister—whether they furnish materials for important secretions, as in the one case, or in the other, as bearers of caloric, are destined merely to keep up the heat of a part.

In the substance proper of connecting tissue the phenomena of nutrition are of rather a low order. The simple diffusion of the nutritious fluid exuded from an occasional blood-vessel suffices to keep up the vitality of the elements which compose it. This is confirmed by examination of the structure of a tendon,

Connecting tissue.

Vessels.

or of the cornea.   There are few organs so poorly
supplied with blood-vessels as the former; and in the
latter there are none whatever.   It follows, then, that
the fluid from which they derive their sustenance
penetrates by imbibition into their substance, or that

Nerves. it gets there through a net-work of plasmatic cells.
There are very few nerves which properly belong to
connecting tissue; it is true that some of its membranous expansions contain a large number, but they
do not minister either to the nutrition, or to the
general sensibility, of the tissue which surrounds them.
The comparative study of nervous distribution in tendons, and in certain regions of the skin confirms this
view.

We may conclude then that, for its own use, connecting tissue is but scantily supplied with blood-
vessels and nerves; it bears to them mainly the relation of a mechanical support.

Distribution. Connecting tissue is distributed very generally and
universally throughout the body, either collected in
bundles or fasciculi of fibres, or spread out in the form
of a membranous expansion.   It serves as the bond
of union between the several parts of an organ, and
it maintains entire organs in their proper relations to
each other.   Alone, it constitutes tendons, ligaments,
aponeuroses, periosteum, perichondrium, the dura-
mater, pia-mater, and sclerotica.   As membrane, invested with epithelium, it constitutes the serous, synovial, and mucous membranes, as well as the skin,
and the membranous expansion which forms the basis
of most glands.

The vitreous humor of the eye, and similar hyaline

and gelatiniform tissues, composed of an amorphous substance, throughout which a variable number of branching or plasmatic cells are distributed, consist, probably, of connecting tissue in an embryonic or imperfectly developed state.

The fibres of connecting tissue develop themselves Development. from cells of the simplest form, which commence the process by assuming an elongated shape, then join each other—end to end, and gradually break up into fibres within, so (see fig. III. pl. IV.) that each row of cells thus attached by their extremities, is developed into a bundle of connective fibres. Whilst the majority of the original cells are thus transforming themselves into connective fibres, others assume a star-shape, sending out branching processes, which, joining themselves to similar prolongations from neighboring cells, produce, after the disappearance of their nuclei, elastic fibres. The cells heretofore called "plasmatic" are nothing more than the star-shaped corpuscles just described, before taking on their final transformation into elastic fibres. This is the mode of development of the fibres of connecting tissue most generally admitted, but we have recently recognised two other modes in which connective fibres are formed, which have never been described, and which perhaps are worthy of notice. A fibrous tumor of the dura-mater, of an encephaloid aspect, presents, in its softer portions a series of oval or fusiform cells, arranged, end to end, in longitudinal rows (Pl. IV. fig. IV. 2). In the harder parts of the same tumor, where to the naked eye it has a distinctly fibrous appearance, its cells are longer and more thread-like, their bodies have become more attenuated, and their nuclei, apparently from increasing compression, have withered and mostly disappeared entirely. Meanwhile these elongated cells, in contact by their extremities, have grown together, and thus become transformed into a continuous fibre. The essential feature in this process, and that in which it differs from the usual mode of development, consists in the fact that there is no tendency observed in the cell contents to break up into fibres, so that each row of cells is eventually fused into but one solitary fibre, and not into a bundle of fibrillæ; (Pl. IV. fig. IV. 3).

In the other mode of development of connecting tissue, which we had an opportunity of studying in a fibrous tumor of the uterus, the formation of the fibre seemed to be due to the metamorphosis of free nuclei. At certain points of the tumor an agglomeration of oval, or spherical nuclei was observed, imbedded in a soft amorphous substance (blastema).* The outlines of these nuclei were clear and well marked; their contents were composed of very fine granules, in some instances so grouped as to represent a nucleolus. Very rarely an outline representing the trace of a cell-wall could be recognised. They measured about $\frac{1}{100}$th of a line in diameter. (Pl. IV. fig. V).

At other points these nuclei were seen more elongated in their shape, stretching themselves, as it were, in the blastema in which they were imbedded, and becoming connected together by their extremities so as to form one fibre out of each row of elongated

---

* This term, originally introduced by Schwann, signifies a soft, solid, hyaline or amorphous material—such as that in which cell growth usually takes its origin.—(*Ed.*)

nuclei (Pl. IV. figs. VI. and VII.). During this process of development of the nuclei the blastema underwent no change, and showed no disposition to take on a fibrous arrangement; it simply diminished in quantity.

Finally, some observers have asserted that simple amorphous substance, presenting no trace of organization, was capable of undergoing spontaneous transformation into connecting tissue. Thus far we have failed to recognise this mode of development, and, moreover, are indisposed to admit the possibility of the spontaneous generation of a recognisable tissue in an entirely unorganized and amorphous mass of substance.

CHAPTER III.

Cartilage—Bone—Teeth.

Sect. I. Cartilage.—The tissue known as cartilage consists of cells of characteristic appearance imbedded in a peculiar material—which we shall designate as the *fundamental substance* of cartilage. This latter exists in two forms, the one entirely structureless, the other distinctly fibrous; hence the two varieties of the tissue: true cartilage and fibro-cartilage.

Cartilage cells. The perfectly formed cartilage cell, such as we find, for example, in the interior of an adult costal cartilage, is usually spherical, or many-sided, and of considerable volume ($\frac{1}{80}$th—$\frac{1}{60}$th of a line). It consists of an external envelope with transparent granular contents, presenting no peculiar features; but the nucleus is filled with large sized oil-globules to such an extent that no nucleolus is distinguishable (Pl. V. fig. I). Sometimes even the granular contents of the cell are replaced by these oil-globules so completely as to give it the appearance of a vesicle filled by a drop of oil. In childhood, and on the surface of the cartilages of the adult, the cells are of smaller size and elongated in shape, and contain very little free fat, especially in the fœtus.

But the distinctive characteristic of the cartilage cell is the existence of a structureless membrane, or capsule, which surrounds it completely on all sides, and which is continuous by its external surface with

the fundamental substance of the tissue. Sometimes
this capsule includes but one cell, and this we see in
examining a cartilage near its surface (Pl. V. fig. II);
more frequently, however, it encloses several, but
rarely more than five or six (Pl. V. fig. I).

The fundamental substance of true cartilage is a Fundamental substance.
hard and elastic material in which no trace of struc-
ture can be detected. In old age, and sometimes
even in adult life, it becomes infiltrated with fat, and
often presents minute cracks—which appearance has
been mistaken for the spontaneous generation of fibres
in a structureless material. But in reality they are
no more fibres than the granular striæ of fibrine—to
which they bear an accurate resemblance.

This fatty transformation, or atrophy, is often found
in the costal cartilages, and is recognisable by the
naked eye in the form of dead white or reddish yellow spots.

Cartilage is made up exclusively of the elements Cartilaginous tissue.
just described. In adult life neither nerves nor blood-
vessels can be recognised in it. The latter, it is true,
are occasionally encountered, but only during the
forming stage of the tissue, or where it is undergoing
transformation into bone, as we shall see hereafter.
Thus, to sum up in a word, true cartilage consists of
a structureless fundamental substance or basis, studded
with minute cavities, lined by a membrane, and enclosing cells.

Cartilages are enveloped by a membrane called Perichondrium.
perichondrium. This membrane is formed by an
interlacement of connective fibres with delicate elastic
fibres, an occasional nervous fibrilla, vessels in vari-

able quantity, and plasmatic cells. These latter exist in greatest number in the deepest portions of the membrane, and it is to be noticed that those in immediate contact with the surface of the cartilage are not distinguishable in appearance from cartilage cells (Pl. V. fig. II). Are we not justified in concluding from this fact that the growth of cartilage is effected by the transformation of these plasmatic cells of its perichondrium?

Relations between cartilage and bone.

Cartilages are united to bones by immediate contact or apposition; there is no substance or tissue interposed between them. Their opposed surfaces are rough, and their minute elevations and depressions fit into each other accurately.

It was supposed for a long time that the free surfaces of articular cartilage were invested with synovial membrane. Careful examination of the surface

Articular cartilage.

of the cartilage demonstrates, however, that within the cavity of a joint it is entirely naked; it is not even covered by the epithelial layer of the synovial membrane. It is incorrect therefore to describe synovial membranes as shut sacs, and as lining the whole interior surface of an articular cavity. A synovial membrane simply covers the internal surface of the capsular expansion which surrounds the joint, or, to be more exact, it is nothing more than the capsule of the joint covered internally by a layer of epithelium. It is to be understood, however, that those ligaments which present a free surface in the cavity of a joint are also invested by epithelium.

Distribution.

Under the head of true cartilage are included: the cartilaginous skeleton of the fœtus, the costal cartilages, those of the joints, the cartilages of the nose, the thyroid, cricoid, and arytenoid cartilages, and the cartilaginous rings of the trachea and bronchial tubes.

Fibro-cartilage differs from true cartilage only in *Fibro-cartilage.* the nature of its fundamental substance or basis, which, instead of being structureless, is fibrous. The fibres of which it consists are of the elastic variety, at least in most fibro-cartilages (Pl. V. fig. III). The intervertebral discs and the semi-lunar cartilages of the knee-joint we have found to be the only exceptions to this rule; their fundamental substance consisting almost entirely of connective fibres.

The principal fibro-cartilages are those of the *Distribution.* external ear and Eustachian tube, the epiglottis, the little cartilaginous masses at the summits of the arytenoid cartilages, the intervertebral discs, and inter-articular cartilages.

Cartilage, like all other tissues, is developed from *Development.* embryonic cells. Those cells which are about to take on the cartilaginous transformation, secrete from their external surfaces an enveloping membrane, which becomes their capsule, whilst a solid structureless material is deposited around them, constituting the fundamental substance. In the formation of fibro-cartilage a portion only of the original formative cells take on the changes above described, whilst the remainder transform themselves into connective and elastic fibres.

The growth of cartilage is effected in part by the *Growth.* endogenous multiplication of its cells (*vide* sect. 1, chap. 1), and in part by the addition of new tissue to its surface, derived from the plasmatic cells of the perichondrium as already described. It has not been demonstrated that cartilage is ever reproduced when destroyed by disease or injury; it is replaced by transformed plasmatic cells, as far as the process can be traced.*

To study the structure of cartilage very thin slices should be cut from it by means of a razor.

SECT. II. BONE.—To study advantageously the minute anatomy of bone, sections as thin and delicate as possible should be made with a saw in every direction through its substance, and the laminæ thus procured should be rubbed down with moistened pumice stone, and afterwards polished upon a fine whetstone. It is well also to examine, in connexion with these, the delicate and transparent scales which form the walls of the cancelli of the spongy portion of the bony tissue; they can be readily detached, and when placed between two slips of glass, and moistened with a drop of water, are ready for use.

Structure of bone.

In a transparent lamina of bone thus prepared, when placed under the microscope, there are always two elements to be recognised, and these alone constitute true osseous tissue, viz. *bone cells*, and the

* There are many points connected with the histology of cartilage, and especially of articular cartilage, still unsettled, and in consequence of the important bearing of this knowledge upon the principles of surgery, as applied to diseases of joints, I cannot refrain from calling attention to the admirable papers of Mr. R. Barwell, F.R.C.S.E., Ast. Surgeon Charing Cross Hosp'l, Lond., in the No. for October, 1859, of the *Medico-Chirurgical Review*, and in the No. for February, 1860, of the *Edinburgh Medical Journal*. They are complementary to the researches of Ecker, Goodsir, and Redfern, and comprise the fullest knowledge of the subject yet acquired by science.—(*Ed.*)

medium in which they are found, which we shall designate as the *fundamental substance* of bone. This latter consists of a whitish structureless material— opaque, or transparent, according to the thickness of the section. It is composed, chemically, of earthy salts, and an organic substance by means of which the earthy particles are held together.

The cells of bone (called also osseous corpuscles, osteo-plastic cells, and *lacunæ*) bear some resemblance in their shape and outline to the star-shaped or branching plasmatic cells already described. They are minute fusiform bodies, slightly flattened laterally, and measuring from $\frac{1}{188}$th to $\frac{1}{118}$d of a line in length. From their exterior a delicate tracery of minute thread-like prolongations radiate in every direction, anastomosing with each other, and with those of neighboring cells. Under a magnifying power of from 350 to 400 diameters it can be distinctly seen that these filiform appendages of the cells of bone are hollow in their interior, and are, in fact, very minute tubes or *canaliculi;* their mode of communication is likewise very apparent, (Pl. V. fig. IV). The more delicate scales of the spongy variety of bone, and the *cementum* of the teeth, present in fact no other constituent elements; but this is not true of the more dense or cortical substance of bone, or of scales of greater thickness. On placing a transverse section of a long bone under the microscope, it is at once apparent that its cells are grouped after a certain fixed plan. In fact they are arranged very regularly in concentric circles around a larger central opening—which is the transverse section of the track of a blood-vessel, or in other words, a Haversian canal. Very many of the canaliculi from the nearest circle of bone-cells are also to be seen running into the Haversian canal. In the long bones the Haversian canals run parallel with the axis of the shaft of the bone, and communicate with each other at short intervals by transverse anastomotic branches (Pl. VI. fig. I. 1, 2, 3). In the short and flat bones they also pursue a determinate course, and anastomose in a similar manner. These canals, which contain the bloodvessels of the osseous tissue, tunnel its fundamental substance in all directions, terminating either upon the external surface of the bone, or in its medullary cavities. The nutrition of bone is effected by means of the parts just described. The very numerous orifices of the canaliculi in the walls of the Haversian canal receive the nutritious fluid which exudes through the walls of its contained bloodvessel, and convey it throughout the network which they and their parent bone-cells or lacunæ form, to the outermost of the series of concentric circles.

Periosteum. The fibrous membrane which invests bone externally, called *periosteum*; resembles perichondrium in its structure ; it is an interlacement, or rather a felting, of connective and elastic fibres, traversed by some nerves and very numerous bloodvessels, and studded with plasmatic cells which, as we shall see, play an important part in the formation and growth of bone. (Pl. VI. fig. V.)

Medullary cavities. The medullary cavities of bones are filled by marrow, which is in direct contact with their walls, for the prevalent idea that the walls of these cavities are lined by an internal periosteum, or medullary mem-

brane, is incorrect. These names have been applied
to the scattered fasciculi of connecting tissue by which
the blood-vessels and fat cells of the marrow are supported.

Marrow is found only in the cancelli and medul- Marrow.
lary canals of bone; neither the Haversian canals
nor the canaliculi contain it. In the fœtus it is of a
reddish color and possesses some consistency; in the
adult it is met with in this form only in the smaller
cancelli of spongy bone, and in short and flat bones;
in the medullary canals of long bones, and in the
larger cancelli of their spongy substance, it is yellow
in color and almost diffluent. Thus, there are two
varieties of marrow, which differ in their histological elements as well as in their physical properties.
The red, or fœtal marrow, is made up of an aggregation of spherical cells, each containing fine granular
matter and one large nucleus. Some of these cells have
several nuclei, and attain a large size ($\frac{1}{17}$th to $\frac{1}{15}$th
of a line). It is worthy of remark, in passing, that
the cells of fœtal marrow are identical in appearance
with certain forms of so-called cancer cells. This
variety of marrow is richly supplied with blood-vessels, which traverse its substance, accompanied by
delicate filaments of connecting tissue.

The cells of yellow marrow are nothing more than
vesicles filled with liquid fat, or ordinary fat cells.
In some of them the nucleus can be still recognised,
and others again resemble so closely the cells of fœtal
marrow as to suggest a series of transitional changes,
by which it is rendered probable that the ordinary
yellow marrow is nothing more than fœtal marrow,

the cells of which have undergone the process of fatty degeneration.

Of the two varieties the red, or fœtal marrow, is more richly supplied with blood-vessels.

Arteries and nerves of bone.

The arteries of bone are derived from its periosteum; one class of them, the smaller vessels, penetrate the compact substance and pursue the same course as the Haversian canals which they occupy; the other class, larger, and known as nutritious arteries, enter separate canals of their own, and thus reach the medullary cavities, where they terminate by supplying the marrow and anastomosing with the vessels of the first class. The veins, as a rule, present the same calibre and pursue the same course as the arteries with which they correspond; in some instances, however, they assume a larger size and different arrangement, as in the sinuses of the diploe, and of the bodies of the vertebræ. Up to the present time lymphatics have not been demonstrated in bone. Its nerves, which are numerous, ordinarily follow the course of the arteries, and supply the marrow as well as the bone; before penetrating its substance they give off branches to the periosteum. There is reason to believe that they terminate by free extremities.

Development of bone.

The development and growth of bone is accomplished in two ways: by ossification of the cartilaginous skeleton of the fœtus, and by transformation of the deeper layers of the periosteum.

The first mode of development is best studied in very thin sections of a young bone, made just at the line of junction of the cartilage and bone. On examining, in the first place, the cartilaginous portion of the section, it is to be observed that its cells are disposed in parallel rows, and that some of them are quite altered in appearance. One portion of them differs in no respect from ordinary cartilage cells, whilst the remainder have already changed in form, the change being confined mainly to their nuclei. The nucleus, for example, has become very irregular in its outline, and by sending out prolongations in every direction, has put on a decided resemblance to a bone-cell. It is imbedded in a finely granular substance limited by a pale circular or oval line, the cell wall; outside of this another line is to be seen in close proximity to the first, and surrounding the cell; this is the capsule of the cartilage cell. (Pl. VI. fig. III. 2, 3, 4.)

In view of these facts it is pretty evident that the osseous cell, or lacuna, is identical with the nucleus of the original cartilage cell, in a more advanced stage of development.

The process of ossification is completed by the elongation of the filiform prolongations, or canaliculi, given off from the nuclei of the cartilage cells, which terminate by anastomosing with the canaliculi of neighboring nuclei; and meanwhile earthy salts have been brought by the bloodvessels and deposited in the fundamental substance of the cartilage, as well as in the interior of its cells. Neither the walls of these cells, nor their enveloping capsule, disappear immediately after the ossification of their contents; by the addition of dilute hydrochloric acid to a portion of recently ossified bone they can both be rendered visible—prssenting their usual appearance.

It is asserted by some observers that it is the cartilage cells themselves, and not their nuclei, which are thus transformed into bone cells. They describe the cell-wall as becoming wrinkled, and undergoing the changes which we have attributed to its nucleus, whilst the nucleus itself fades away and finally disappears entirely. (Pl. VII. fig. II.) In opposition to the authorities by whom this statement is endorsed, we are disposed to persist in the belief that it is the nucleus of the cartilage cell which becomes transformed into the osseous cell, or lacuna, of bone.

Fœtal Marrow. The cells of ossifying cartilage which remain unchanged during the process just described, eventually assume the character of fœtal marrow. At first they become the seat of active endogenous development, in consequence of which they increase, together with their enveloping capsules, very considerably, in size. Pl. VI. fig. IV. 4.) Very soon they come in contact with each other, their capsules becoming welded together (Pl. IV. 5), and the partitions which thus result, melting away as they lie arranged in longitudinal rows, a medullary canal is thus formed filled with young cells and free oil-globules, which, in fact, constitutes fœtal marrow. (Pl. IV. 6.) Whilst the process of ossification is going on, numerous blood-vessels derived from its perichondrium are seen ramifying through the substance of the cartilage. These seem at first to be simply hollowed out of the cartilage; but later in the process the cartilage cells immediately around them become elongated and fusiform in shape, and seem by their subsequent union to form walls for the vessels.

The process of ossification from periosteum is less complicated than that just described, by which cartilage is converted into bone. This fibro-vascular membrane contains, as we have already stated, a large number of little star-shaped or plasmatic cells. When, by the addition of acetic acid, the ordinary connective fibres of the membrane are made to disappear, more than one layer, composed of elastic fibres and branching cells, is brought in view, in which both the form and arrangement of osseous cells are faithfully represented. (Pl. VII. fig. 1, 3.) In the deepest layers of the periosteum, where ossification takes place, the plasmatic cells are seen to be more numerous, and farther advanced in development than elsewhere. The blastema in which they are imbedded is also deeper in color, and this is explained by the presence, already, of earthy salts. (Pl. VI. fig. V. 2, 3.) In fact the process of ossification is thus almost perfected, for the plasmatic cell requires only to be imbedded in and surrounded by earthy matter, to become a cell of bone. However, the process is not always effected in so simple a manner, for sometimes the plasmatic cells do not present their usual star-shaped prolongations, and in this case the filiform processes, which ultimately form canaliculi, only make their appearance whilst the incrustation of their cell-walls is actually taking place.

From our own investigations it is evident that the ossification of the cranial bones is effected entirely by the changes just delineated in their periosteal coverings. (Pl. VII. fig. I.) The new deposits of bony matter which take place in some grades of perios-

teal inflammation are explained in the same manner.

The reparation of bones after fracture or exsec-
tion, or where a portion has been gouged out in an
operation, is accomplished by a process analogous to
that which the periosteum performs in ossification.
The gelatiniform mass which forms between the fragments of a broken bone, in an excavation produced
by the gouge, or even in a medullary cavity, contains
usually some fibrillæ of connective tissue, together
with a large number of blood-globules and oval
nuclei (fibro-plastic), which are subsequently converted into bone cells. It is easy to follow the successive transformations of these nuclei by the microscopic examination of a very thin lamella of bone to
which this gelatiniform material is still adherent.
These are the several steps of the process : at a little
distance from the bone, the oval nuclei possess a very
regular outline, but as we trace them nearer to its
surface they are observed to change their shape;
their outlines wrinkle, and send out linear prolongations radiating in every direction; at the same time
earthy matter is deposited around them, by which
they become gradually encrusted, and thus the metamorphosis into bone is completed. (Pl. XXVII.
fig. I.)

Reparation or reproduction of bone in a medullary
canal, and in the cancelli of its spongy portions, is
not accomplished by medullary membrane, which, as
we have already asserted, has no existence, nor yet
by the aid of an imaginary cartilage, which, in any
case, would be but transitory. The phenomena of

ossification here are entirely analogous to those we have observed in the deeper strata of periosteum, both in its healthy state, and when inflamed. In all these cases the osseous cell, in the absence of which the tissue of bone cannot exist, is developed from another cell or its nucleus—that is to say from the essential organic element—which is always present both in the periosteum, and in the contents of the medullary cavities.

SECT. III.—TEETH. A tooth consists of a central Teeth. portion which constitutes nearly the whole of the organ, and of an outer lamina which accurately invests its external surface. The central mass, or *ivory*, has a cavity in its interior—variable in size and shape, which communicates externally through a narrow canal situated at the extremity of the root; this cavity contains the *pulp* (Pl. VIII, fig. I. 1, 2). The portion of the external lamina which invests the crown of the tooth is the *enamel* (fig. I. 4); that which covers its root is the *cementum* (fig. I. 3).

The *ivory* is composed of fundamental substance, Ivory. traversed by minute canals. The former, structureless and transparent when viewed in thin section, is identical with the fundamental substance of bone, and the canals which it contains resemble closely the canaliculi of bone-cells. They take their origin in the central cavity of the tooth, and thence radiate to the surface of the ivory, where they terminate; it is an exception to the rule for them to pass beyond this limit and to penetrate the exterior lamina (Pl. VIII. fig. III. 2). At their commencement some are single, others arise from a common trunk, and others again take their origin from minute cavities in the deeper strata of the fundamental substance; these latter, however, always communicate with the central cavity

Canaliculi of the ivory.

of the tooth (Pl. VIII. fig. II. 4). With a magnifying power of 350 to 400 diameters, the canals can be seen to be sharply limited by two very fine but distinct lines, to be slightly wavy or undulating in their course, and to run, mainly, parallel to each other. Their lateral branches can also be distinguished, radiating in every direction, anastomosing with each other, and thus forming an universal network, which permeates, everywhere, the fundamental substance of the ivory. (Pl. VII. fig. III. 1, 2 ; Pl. VIII. fig. II).

At their origin the canals are larger than at their termination, measuring, on an average, from $\frac{1}{1300}$th to $\frac{1}{700}$th of a line. Sometimes they present, in their course, small spindle-shaped enlargements (Pl. VIII. fig. II. 3), and they terminate, as a rule, in irregularly shaped cavities which are the interspaces between minute globular masses of the fundamental substance (Pl. VIII. fig. II. 5). It is to be noticed that these inter-globular spaces, as they have been designated, communicate freely with the bone-cells of the *cementum*, to which, in fact, they bear no little resemblance.

Enamel.

The *enamel* forms a hard and homogeneous lamina moulded accurately upon the surface of the crown of the tooth, and terminating, at its neck, by a thin edge, which seems to insinuate itself between the ivory and the terminal margin of the *cementum*. It is composed exclusively of five or six-sided prisms,

placed side by side, without any appreciable intervening substance (Pl. VIII. fig. IV. 1). Examined
in vertical section they are observed to undulate
slightly, and to assume a direction perpendicular to
the nearest surface; moreover, they are generally
parallel with each other, except upon the irregularly
shaped surfaces of the molar teeth, where they form
divergent bundles (Pl. VIII. fig. III.). Their substance is wholly amorphous; sometimes it is crossed
by transverse lines; its chemical nature seems to
connect it with the epithelial formations. Some
observers have asserted that the enamel is invested
externally by a delicate structureless layer, which
they designate as the cuticle of the enamel.

The *cementum,* by which the root of the tooth is Cementum.
covered externally, is true bone; it consists of fundamental substance, and bone-cells of variable size and
irregular in their arrangement. Haversian canals are
absent, except where its structure has been altered
by inflammation. The periosteal investment of the
alveolar sockets also covers the surface of the cementum.

The *pulp* of the tooth, which occupies its central Dental pulp.
cavity, is connected with the periosteum of the socket
by a pedicle which penetrates the orifice at the
pointed extremity of its root. It is made up of delicate connective tissue, interspersed with plasmatic
cells, and largely supplied by blood-vessels and
nerves—the terminal arrangement of the latter being,
as yet, imperfectly made out. The presence of lym-
phatic vessels in the pulp is uncertain. During the Development of the teeth.
sixth week of foetal life, the free borders of the max-

illary bones begin to show distinct longitudinal
depressions or grooves, at the bottom of which minute
granulations make their appearance, which are the
germs of the teeth; it is from these that the ivory is
developed. Shortly afterwards partitions make their
appearance by which the groove is divided up into
separate apartments, with a germ at the bottom of
each—recalling in their appearance the *circumvallate
papillæ* of the tongue. Still later the margins of
these compartments, as they grow, rise above the
level of the germs, contract by approaching from
opposite sides, and finally unite together. Hereafter
the germ is enveloped on all sides in a cavity which
takes the name of the *dental sac.*

Dental sac.   The walls of the sac are at first formed by two distinct layers, which afterwards become one. The
external layer, which later becomes the periosteum
of the socket, is made up of very vascular connecting
tissue; the internal layer, of the same nature as the
preceding, but more delicate in structure, contri-
butes, according to M. Magitot,* to the subsequent
formation of the enamel.

We know already that the dental germ takes its
origin, by a pedunculated root, from the bottom of
the sac; from a point diametrically opposite to
this springs another germ similar to it in nature,
which, as we shall see hereafter, gives origin to the
enamel.

Development of
the ivory.   The dental germ (whence the ivory is developed)

* Emile Magitot has a paper on the structure of the teeth in the
Archives Medicales, Paris, Jan. 1858, and has since written on the subject
in same Journal.—(*Ed.*)

rich in blood-vessels, which reach it through its pedicle, contains, besides, a large number of nuclei and young cells of an oval·form, together with some connective fibrillæ; its nerves, which appear somewhat later, accompany its vessels. It is covered by an amorphous membrane* which, as Magitot has shown, is incorrectly regarded as a portion of the internal layer of the wall of the dental sac. This border belongs really to the germ, and takes no part in the subsequent formation of the ivory. Beneath this there is a stratum of oval cells, arranged very regularly side by side, with their long diameters perpendicular to the surface of the germ. The peripheral extremity of each cell elongates into a thread-like tube, which by degrees increases in size, and gives off minute lateral branches; in this manner a great number of minute canals are formed, which run parallel with each other to the extreme limit of the germ, communicating largely by their ramifications. Thus the canaliculi of the ivory of the tooth are developed from the superficial cells of its germ, each cell, according to Kölliker,† by a process similar to wire-drawing, elongating itself into a complete canal. Whilst the dental canals are being thus developed,

* Homogeneous basement membrane of Todd and Bowman. In his account of the development of the teeth the author mainly follows Goodsir.—(Ed.)

† A. Kölliker, Professor of Anatomy and Physiology in the University of Würzburg (Bavaria), author of the "Manual of Human Microscopic Anatomy," which has been twice translated into English; first by Professors Busk and Huxley, and published by the Sydenham Society, 1853–54; and more recently, in a revised edition, by Dr. George Buchanan, under the author's supervision. Parker, London. 1860. —(Ed.)

an amorphous substance is poured out in the intervals between them, an exudation furnished, no doubt, by the deeper cells of the germ, and which subsequently becomes the fundamental substance of the ivory of the tooth. Finally, its development is completed by the infiltration of the fundamental substance with calcareous salts, and what remains of the original germ becomes atrophied and forms the dental pulp. The germ of the enamel, having attained its complete development, envelopes entirely the base of the germ of the ivory, which we have traced to its complete ossification. Superficially it is made up of connecting tissue rich in blood-vessels; more deeply there are only star-shaped cells imbedded in amorphous material; and finally the stratum immediately in contact with the dental germ, is formed by an epithelial layer, whose cells, long, narrow, and prismatic in shape, resemble closely the prisms, already described, of the enamel. Thus it is more than probable that the enamel is formed directly by the petrifaction of these elementary bodies. The lower partions of the dental sac give origin to the cementum, taking on the process of ossification in the same manner as periosteum.

In view of the fact that connecting tissue, under certain circumstances, undergoes transformation into cartilage and bone, as well as into the substance of the teeth, these tissues, being all analogous in nature, are grouped together, by some authorities, in one family.

The preparations required for the study of the structure of the teeth are prepared in the same manner as sections of bone.

CHAPTER IV.

Muscular Tissue.

THE essential element of muscular tissue, *i.e.* the contractile element, presents, in its intimate structure, both cells and fibres. The cell is a transitory element, or rather is found only in those organs in which an incomplete stage of development is persistent. The fibre, in two distinct forms, constitutes the bulk of all recognised muscles. The two varieties are known as the striped, and the smooth or unstriped muscular fibre.

Smooth, or unstriped muscle.—When a small portion of certain muscular organs, the muscular wall of the intestine, or of the bladder, for example, is placed beneath the microscope, it is found to consist apparently of long, pale, spindle-shaped bodies, each one of which is provided with an elongated nucleus having a clear and distinctly marked outline, and surrounded by a substance so finely granular as to appear almost amorphous. More attentive examination, however, reveals the real nature of these nucleated bodies—the essential contractile elements; in place of a spindle-shaped cell, it is soon recognised as a true fibre, presenting, as we trace it in the direction of its length, a regular succession of contractions and enlargements. These fibres, instead of running parallel with each other, cross at very acute angles, and in such a manner that their points of intersection

Structure of smooth muscular fibre.

seem to correspond always with their narrow or contracted intervals, or, as it is still described by some,
with the tapering points of the fibre-cells. A cir-
cumstance which has doubtless served to perpetuate
this erroneous view is that these fibres break very
easily when an attempt is made to isolate them, and
as the rupture always takes place at one of the constricted portions of the fibre, which tapers off as it
breaks, in consequence of its elasticity, it results that
each fragment thus detached presents a regular spin-
dle shape. (Pl. XIV. fig. VIII.)   But if a section is
made across the course of the fibres, polygons are
brought into view, varying very considerably in dia-
meter ($\frac{1}{700}$th to $\frac{1}{160}$th of a line), but never of a size
so small as to be recognisable as the terminal point
of a spindle-shaped corpuscle. (Pl. IX. fig. II.)   The
same section shows also the mode in which these
fibres are grouped together so as to form fasciculi of
muscle.   Between the fibres which are thus in immediate contact with each other, forming fasciculi, there
is no intervening substance whatever ; but they are
surrounded by an investment, or sheath, of connecting
tissue which also includes the blood-vessels and nervous filaments by which they are supplied. (Pl. IX.
fig. II. 4.)

Contractile cells.   The contractile fusiform element, or fibro-cellular
corpuscle, is found only in those organs whose nor-
mal condition is one of imperfect development—as
for example, in the smallest arteries, of a diameter of
from $\frac{1}{70}$th to $\frac{1}{40}$th of a line.   In vessels of this class
the middle or muscular coat consists of this element
in its purest form.   It is not difficult to demonstrate

that the contractile element of which it is composed is a spindle-shaped, or fusiform, corpuscle of variable length, in the interior of which a nucleus, which approaches more nearly the circular shape than that of the true smooth or unstriped fibre, is recognisable. (Pl. XV. fig. VII. 2, 3.) We find this fusiform corpuscle in the walls of the villi of the small intestine, in the muscles of the hair-bulbs, and perhaps also in other organs. It is as yet an unsettled question whether the *dartos* is made up of this element, or of the fully developed unstriped fibre.

The distribution of the smooth, or unstriped muscular tissue, limited at first to organs possessed of obvious contractility, is daily increasing in extent. Thus, its presence has been demonstrated in the *villi* of the intestines, in the excretory ducts of most of the glands, in the middle coat of arteries, veins, and lymphatics, in the genital organs of the female (*uterus* and appendages, *vagina*, and *corpora cavernosa* of the *clitoris*); in the genital organs of the male (*corpora cavernosa penis*, prepuce, prostate, seminal vesicles, &c.); in the vascular tunic of the eye; and, finally, throughout the whole extent of the skin, where it is very unequally distributed. It is found in connexion with the hair-bulbs and sebaceous follicles; and by its presence, the phenomenon of horripilation, or goose-flesh, is explained. Some portions of the external integument are unusually rich in unstriped muscular fibres; for example, the prepuce, and the skin of the nipple, and in some instances, of the whole female breast. The elongation and rigidity of the nipple are due entirely to their contraction, and not,

Distribution of smooth muscular fibres.

as generally supposed, to a turgescence, or erection, similar to that of the *corpora cavernosa*.

Development of unstriped muscular fibre. Smooth, or unstriped muscular fibres are developed from formative cells, which at first elongate, and then become attached by their extremities. In some organs and localities this mode of development is incomplete; the metamorphoses of the formative cells are arrested in their earlier stages, and thus the existence of the contractile fusiform corpuscle—the fibre-cell of Kölliker, is explained. Hereafter we shall see that the true striped muscular fibre passes through the two stages we have just described before it assumes its ultimate and definite form, so that, in a general histological survey of the entire muscular tissue of the body, it may be considered as representing, in its several constituent portions, different and progressive degrees of development of the same formative element—of which the striped fibre is the last and most perfect form.

In the muscular tissue of the uterus during pregnancy we recognise the development of new muscular fibre of the smooth variety. According to Kölliker this takes place only during the first six months of gestation. In its deeper strata an immense number of cells is recognisable, measuring from $\frac{1}{1000}$th to $\frac{1}{50}$th of a line, and we can trace them through the several phases of their development into smooth muscular fibres. After delivery, and whilst the uterus is diminishing in bulk, the larger proportion of these muscular fibres become infiltrated with fat, break down, and disappear entirely by absorption; it is by this process of fatty atrophy that the uterus returns, after child-birth, to its original dimensions.

Striped Muscle.—The tissue of muscle in its simplest element—the primitive fibre—is variable in its physiognomy, presenting to the eye but two constant features, viz: an external envelope, with contents marked by transverse stripes. (Pl. IX. fig. V. and Pl. X.)

The primitive fibre is usually polygonal, rarely cylindrical; its envelope, called myolemma or sarcolemma,* is readily recognisable, either without any previous preparation, or by the aid of chemical reagents, as a perfectly structureless membrane. Neither in the living nor dead fibre, neither in the state of contraction nor of relaxation, can any folds or wrinkles be detected in it corresponding with the cross stripings of its contents. On its internal surface, at regular intervals, oval nuclei (Pl. X. fig. II. 3; Pl. IX. fig. V. 3) are visible—the last traces of the cellular origin of muscle. It is exceedingly elastic.

On examining its contents, the most marked features observed are the transverse stripes, equidistant from and parallel with each other. Occasionally longitudinal striæ are to be detected in muscular fibre instead of the transverse stripes, and in rare instances both the longitudinal and transverse stripes are present. (Pl. X. fig. II. 4, 5.) If this contained substance is still farther analysed by the aid of chemical re-agents (chromic acid, alcohol, etc.) or prepared by boiling, and even sometimes without the employment

Striped fibre.

---

* This term was introduced by Bowman, who first investigated and described the sarcolemma, and its relation to the primitive muscular fibre of the striated variety. V. *Cyclop. of Anat. and Physiology*, Art. Muscle, and Todd and Bowman's *Phys. Anat.*—(*Ed.*)

of these artificial means, it becomes evident that it is composed of two distinct constituents—one granular and the other amorphous, the latter being very variable in amount, and serving as a connecting medium to the former—which is thus imbedded in it. (Pl. IX. fig. V. 5.) These minute granules—the *sarcous elements* of Bowman (Pl. X. fig. II. 5), are slightly flattened in the direction of the length and breadth of the fibre, and measure, on an average, $\frac{1}{10000}$th of a line. These little bodies seem to be the active agents in the production of the contractility of muscle; it is in them, and in their relation to the amorphous material by which they are enveloped, that we are to look for the power which the muscular fibre possesses of changing its size and shape.

In fact, according as these ultimate corpuscles approach each other more closely in the direction of the length of the fibre than in the direction of its breadth, it will assume the fibrillated appearance (Pl. X. fig. II. 4, 5), or, in the opposite case, the appearance of discs piled upon one another. This latter disposition of the sarcous elements presents a more striking appearance, when between the discs which result from their arrangement in transverse rows, a large amount of the amorphous hyaline substance is present, as in fig. V. (Pl. IX. 5).

To sum up in a word: the ultimate fibre of striped muscular tissue is composed of an external envelope of simple structureless membrane, which contains a granular material of soft consistence; and the variations in appearance of the striæ by which it is marked, are due to the varying relations which these granules are capable of assuming to each other.

'I he striped muscular fibre is continuous throughout its whole length, and this corresponds accurately with that of the fasciculus, or bundle, of which it forms a constituent part. Up to the present time we know of but two organs which constitute exceptions to this law, and these are the heart and the tongue. The muscular fibres of the heart are branched, and form very frequent anastomoses, recalling the disposition of the *columnæ carneæ* of the ventricles, and explaining the admirable correspondence and unity of action in the movements of the organ (Pl. X. fig. III). The fibres of the tongue give off branches only in its submucous layer of muscular tissue, and these do not appear to anastomose. They terminate by pointed extremities which are attached to minute fasciculi of connective fibres, which seem to act as their tendons —at least this is the opinion of most authorities. The ultimate fibres of striped muscle are united together by delicate lamellæ of connecting tissue (*perymisium*), and thus constitute secondary fasciculi. These again, surrounded by sheaths of the same material, but somewhat more dense, form tertiary fasciculi; and, finally, the entire muscle is enveloped by a still stouter sheath which forms its external *perymisium*. In this outer membrane we find the ramifications of the nerves and nutritious vessels of the muscle.

The striated fibre terminates by a rounded extre- mity, which is simply in close apposition with its tendon. Nevertheless Kölliker asserts that this arrangement, which is certainly the most common of all, is found to exist only where the muscular fibre

meets its tendon obliquely; but where it is continuous, in the same line with the tendinous fasciculus
with which it corresponds, the two tissues blend
together insensibly, without any line of demarcation
at their point of contact.*

Aspect of the striated fibre during contraction.

We cannot close our description of muscular fibre
without remarking that Weber[†] long since demonstrated that it does not wrinkle during contraction.
It becomes shorter by increasing in breadth and
thickness, at the same time, just like a cylinder of
caoutchouc after being stretched. The zigzag wrin-
kling occurs only where the extremities of the fibre, as
it shortens, do not return by the same line in which
it was elongated. It is very easy to verify the exact-
ness of Weber's assertion by examining, under the
microscope, the glosso-laryngeal muscle of the frog,
when under the influence of galvanic stimulus.

Nerves of muscle.

The nerves of striped muscle are supplied both by
the cerebro-spinal axis, and the sympathetic system;
but those derived from the latter source are very few
in number.

* "In the human body the smooth muscular tissue nowhere forms large
isolated muscles, as in the case of the recto-penal muscles of mammals,
for example, but occurs either scattered in the connective tissue, or in
the form of *muscular membranes*. In both cases its bundles are either
arranged parallel to each other, or united to form networks; and, even
in man, are connected in many places with *tendons of elastic tissue*, as
first detected by me in the tracheal muscles, and in the cutaneous feather
muscles of birds." Kölliker, *Manual of Human Microscopic Anat.*
London, 1860, p. 65.—(*Ed.*)

† This is not the celebrated Weber, Professor of Anatomy at Leipzig,
and famous for his researches on the structure of the placenta, the skeleton, joints, etc., but E. Weber, author of an elaborate article, "Muskel-
bewegung," in Wagner's Handwörterbuch, 1846.—(*Ed.*)

As they enter the substance of a muscle, its nerves are united in bundles, and pursue a course almost at right angles to its fibres. Very soon they divide and subdivide, gradually approaching the same direction as the fibres of the muscle, and finally the smaller branches run almost parallel with them (Pl. XI). In their course, the nervous filaments separate sometimes from each other without anastomosing; and, again, they unite after separating, in such a manner as to form loops, and even many series of loops, or a network of anastomoses.

As to the mode in which nervous filaments terminate in muscle, the following is the result of our examination of different muscles of the frog. At the points where the little bundles of nerves separate, the majority of their primitive fibres present a distinct contraction in diameter, a strangulation, to half their previous size. From these contractions two branches usually, sometimes three, take their origin; and these, after running a short distance, undergo a similar contraction and branching. Finally, the terminal fibres taper off quite rapidly, soon show but a single outline, and at last seem to lose themselves on the sarcolemma (Pl. XIV. fig. I).

Can we infer from these facts a similar distribution of the nerves of striated muscle in man? From the researches of Valentin,[*] confirmed in part by Kölliker, it would seem not. Both of these authorities assert that they have seen nervous fibres terminating by loops, in the muscles of the smaller mammiferæ, and in man. Lebert* also has given a representation of nerves terminating by loops, in the muscles of the abdomen, and in the tongue of the frog.

Remak has discovered microscopical nervous ganglia in the auriculo-ventricular septa of the frog's heart, and very rightly ascribes to their influence the persistence of the rhythmical pulsations of the organ, when isolated from the body, as well as the continued regular contractions in the septum itself, when this has been detached from the rest of the heart.

From the observations of Reichert, who has carefully studied the distribution of the nerves of the frog's heart, it would appear that each muscular fibre is in relation with several nervous filaments. Volkman has sought to establish the numerical proportion of large and slender fibres which enter the substance of a striped muscle; he has counted twelve slender fibres in an hundred.

Very little is known of the mode of distribution of its nerves to the non-striated muscular tissue. It may be asserted, probably, with truth, that it is much less richly supplied with nerves than the striated variety; the middle coat of the arteries affords evidence of this.

* H. Lebert, at present Professor of Clinical Medicine in the University of Breslau, author of *Physiologie pathologique*, 2 vols. Paris, 1845; the "*Traité d'Anatomie pathologique*," now in course of publication, and numerous papers in the *Annales des Sciences Naturelles*, and elsewhere. —(*Ed.*)

† Bogislav Reichert, Professor of Anatomy at Berlin, successor to Müller.—(*Ed.*)

‡ Professor of Anatomy at Dorpat, Livonia.—(*Ed.*)

Of the diverse theories put forth as to the mode of development of muscular tissue, we shall notice only that which seems to us most in conformity with ascertained facts.

Muscular tissue is derived originally, like all the other tissues, from the primordial cells of the embryo, which, at first, everywhere the same, undergo a special metamorphosis in order to form each and every histological element.

The embryonic cells which are about to form muscular fibres at first increase in length, and then, coming in contact with each other by their narrowing extremities, finish by uniting together. Afterwards, the partitions formed by these lines of union disappear, and we have, as a result of the process, tubes of structureless membrane presenting constrictions (which mark the line of fusion of the original cells), and intervening expansions, each of which contains a nucleus (Pl. IX. fig. IV. 1, 2, 3). At this period their diameter varies from $\frac{1}{1000}$th to $\frac{1}{100}$th of a line. During the process of transformation the contents of the cells, at first hyaline, become granular, and the granules, which bear some resemblance to oil-globules, assume a regular arrangement either in longitudinal or transverse rows (Pl. IX. fig. IV. 4, 5, 6). Still later, the muscular fibre, thus formed, increases in thickness, assumes a cylindrical shape, and its several characteristic features become more and more obvious; the division of its interior into fibrillæ is noticed, and also the appearance of a large quantity of nuclei, which take their origin by endogenous multiplication. If, at this moment, the extremity of a fibre is examined on

transverse section, it is found that the changes just
described take place from its circumference towards
its centre, very rarely in the opposite direction. Finally, the fibrillæ increase in number, whilst their
newly formed nuclei are absorbed, and the muscular fibre gradually puts on its permanent appearance.

The following is a summary, in two words, of this
process—which seems best borne out by facts: fusion
of embryonic cells, which by their union form a tube,
or myolemma; metamorphosis of its contents into
elementary granules, and their subsequent development into fibrillæ or discs.

During the earlier periods of development and
growth of the fibres of muscle, they are enveloped by
a large number of elementary cells—some oval in
shape, others smaller and round. Of these, the
greater proportion are destined to form the connecting tissue and other parts which constitute a complete muscle ; and the rest, after having contributed,
no doubt, to the growth of the persistent elements of
the tissue, fade and disappear. The muscular fibres
of the heart are developed in accordance with the
same rules, with this exception, that, in place of the
fusion and subsequent transformation of simple cells,
these phenomena are accomplished by means of
branching, or star-shaped cells ; hence arises the
branching character of the fibres of this organ. The
fibres of muscle, having attained their complete
development, remain permanently, for an indefinite
period, in this state, and the metamorphoses which
occasionally occur in their interior, are marked

by a very feeble amount of vital activity. Up to the present time the mode in which muscular fibre is reproduced, has not been satisfactorily made out.*

* Wounds of muscles never heal by means of transversely striated muscular substance; . . . they heal simply by means of a tendinous cicatrix. An instance of new formation of muscular tissue has been seen by Rokitansky in a tumor of the testicle of an individual eighteen years old; and another by Virchow, in an ovarian tumor. Kölliker, op. cit. —(*Ed.*)

CHAPTER V.

Elements of Nervous Tissue.

THE analysis of nervous tissue reduces its anatomical elements to two forms, viz. the fibre, and the cell.

Structure of nerve fibre. The fibres of nerve tissue present variations in the details of their structure; sometimes consisting of a tube of a certain diameter, with envelope and contents perfectly distinct; whilst, at others, on the contrary, the envelope and its contents so blend together as to constitute a simple homogeneous fibre.

The envelope of the nerve tube is entirely structureless, and possessed of a certain amount of elasticity. Immediately in contact with its internal surface we find a soft amorphous substance of an albuminous and fatty nature—this is the nervous marrow, or medullary sheath. In fresh nerve fibres this *medulla* is homogeneous, and constitutes a regular tube; but soon after death it becomes disintegrated, and separates itself into lumpy masses, between and upon which the envelope contracts so as to give the fibre a varicose appearance (Pl. XII. fig. I.). Finally, the central axis of the tube is occupied by a cylinder of amorphous material, of an albuminous nature, more compact and more tenacious than the medulla, which is known by the name of *axis-cylinder* (Pl. XII. fig. I. 7). It is exceedingly difficult to make out the axis-cylinder in fresh nerve tissue ; but occasionally it can be detected protruding from the broken extremity of a nerve tube. In order to render it more distinct, a number of chemical reagents are employed ; amongst these, dilute chromic acid seems to be the most efficacious. In nerves still alive, so to speak, those, for example, of a specimen of muscle still capable of contraction—we can distinguish only the envelope of the tube with uniform or homogeneous contents; the axis-cylinder is not recognizable—hence it has been regarded as the result of *post-mortem*, or artificial change.

To repeat: in a nerve tube we observe, first, two outer parallel lines representing the enveloping membrane; then, within these, two other parallel lines, which mark the limits of the medulla; and finally, in its centre, two more lines, always parallel, which form the outlines of the axis-cylinder, when it is visible (Pl. XII. fig. I. 5, 6, 7). The diameter of these tubes varies from $\frac{1}{113}$d to $\frac{1}{317}$th of a line.

There are other fibres, smaller than those just described, in which but two parallel lines on each side of the tube can be made out—one of which corresponds to the envelope, and the other to its contents. The most minute fibres of all ($\frac{1}{1755}$th to $\frac{1}{1155}$th of a line) are simple solid cylinders, limited by two exterior lines only, in which it is impossible to distinguish the envelope from the material contained within it. In the first of these it has been supposed that the medulla was absent—the envelope and axis-cylinder alone remaining, and that the latter, or most minute of all, was formed by the axis-cylinder alone.

Fine nerve fibres.

This, however, is not, as yet, susceptible of demonstration. Nerve fibres remain entire whilst in the nervous centres, and throughout their whole course; but, at their peripheral extremities, most of them divide into branches. This fact is perfectly made out as far as the nerves of motion are concerned, as we have already seen whilst studying the structure of striped muscle. It is probably true also of the nerves of mucous membranes, and of the external integument.

Terminations of nerve fibres. Their mode of termination, which has been a subject of research for so many minute anatomists, is not yet definitely established for all the tissues. It appears to be susceptible of demonstration that in the ganglia, and other nervous centres, nerve fibres are in contact with nerve cells. It is equally beyond doubt that in the muscles, and in certain regions of the skin and mucous membranes, nerve fibres, after undergoing division, terminate by free extremities (as in muscle), and sometimes by again anastomosing with each other (as in the skin and mucous membrane of the tongue). The experiments of Prof. Cl. Bernard, on the phenomena of recurrent sensibility, tend to prove that a large proportion of sensitive fibres end by forming loops. We shall see hereafter that in the eye, the ear, and the olfactory mucous membrane, they terminate by contact with nerve cells, as in the nervous centres. Finally, in the integument of the palm of the hand and sole of the foot, we find, in fibres of sensation, two peculiar modes of termination, viz. the corpuscles of Pacini and corpuscles of Meissner, which we are about to describe.

Corpuscles of Pacini.

The corpuscles of Pacini* are little ovoid, ‹r rather ellipsoid bodies, attached by one extremity, by means of a very delicate pedicle, to the nerves of the fingers and toes. They consist of a central cavity inclosing the extremity of a nerve fibre, and an external envelope. This latter resembles the cornea in its structure; it is made up of a number of amorphous lamellæ, overlaying each other concentrically, between which we find a great number of plasmatic nuclei disposed in regular series (Pl. XIV. fig. II. 2). The more superficial lamellæ lose themselves on the surface of the pedicle. The central cavity is filled·with a fine granular substance, in which the outlines of very pale cells can sometimes be distinguished. Finally, in the central axis of the cavity is a nerve fibre, which is remarkably pale, and, for this reason, not easy to discover. By tracing it onward it is seen to terminate in a slight bulb (Pl. XIV. fig. II. 4), whilst below, it is found occupying the centre of the pedicle, through which it is continuous with the nervous twig on which the Pacinian corpuscle is situated. Some authors have represented the nervous filament as dividing, in the interior of the cavity of the corpuscle, into two or three terminal branches.

Pacinian corpuscles are found also in other localities than in connexion with the nerves of the fingers and toes. Kölliker has met with them in connexion with the cutaneous nerves of the arm and forearm, on the back of the hand and foot, and in the terminal branches of the internal pudic, inter-

---

* Nuovi organi scoperti nel corpo umano del Dottore Filippo Pacini, Pistoja, 1840.—(Ed.)

costal, and sacral nerves, and, finally, in the great sympathetic plexus which surrounds the abdominal aorta.

Corpuscle of Meissner.

The corpuscle of Meissner,* or tactile corpuscle, is a minute microscopical organ which occupies the interior of some of the papillæ of the true skin. To get a view of them, it is necessary to make very delicate sections of the skin of the palmar surface of the last phalanges of the fingers or toes, and to apply dilute acetic acid to the section under the microscope. Its shape, like that of the Pacinian corpuscle, resembles an ellipse (Pl. XXIII. fig. II. 6). Its structure consists of a faintly striated substance studded with plasmatic nuclei (fig. II. 9) arranged transversely. At the inferior, or attached extremity of the corpuscle, a nervous filament is seen applying itself to its surface, in the inequalities of which it seems to lose itself by describing tortuous curves which are only occasionally visible. Whether it ends by becoming continuous with the substance of the corpuscle, or by simple division, or by forming a loop, has not as yet been made out.

Kölliker has found these tactile corpuscles in the papillæ of the vermilion borders of the lips, in the fungiform papillæ of the point of the tongue, on the nipple, the glans penis and clitoris; but they are encountered in greatest number in the skin covering the last row of phalanges of the fingers and toes.

Nervous cells.

Nerve cells, or corpuscles, vary exceedingly both in size and shape. In all of their forms they present, however, every element of a perfect cell. Thus they have a very delicate cell-wall, so thin and delicate, in

---

* Successor to Rudolf Wagner in the chair of Physiology at Gottingen.—(*Ed.*)

fact, as to have led some to doubt its existence. They contain a pale, finely granular substance, together with a variable amount of pigment, as a general rule (Pl. XIII. fig. I. 6). The nucleus is spherical in shape, and much more distinctly marked in its outline than the cell itself, and amongst the granules which it contains the nucleolus is readily distinguished by an unusual degree of brilliancy. In the ganglia there are some cells which, outside of their own proper cell-wall, are enveloped by another, a second and much thicker envelope, consisting of structureless or very delicately fibrillated material, full of oval nuclei (Pl. XIII. fig. I. 8). This seems to be partially developed connecting tissue belonging to the proper stroma of the organ.

In regard to their form, nerve cells are either Shape of nerve cells. simply spherical, or furnished with one, two, three or more, tubular prolongations of their cell walls (Pl. XII. fig. VI.; Pl. XIII. fig. I). The spherical cells seem to be merely in contact with the neighboring nervous elements amongst which they are situated; but the multipolar or caudate cells are continuous, by their prolongations, with nerve fibres, or their caudate processes anastomose with each other, as can be readily seen in the grey matter of the cerebellum.

Nerve cells, associated with other elements, are Distribution. found in the grey matter of the cerebro-spinal axis, and in the ganglia of the cerebro-spinal nerves and those of the great sympathetic. They are also found in the nerves which traverse the substance of órgans, in which they form microscopic ganglia. Finally, as we have already intimated, the nervous fibres of the retina terminate in globular masses, and likewise those of the internal ear, and olfactory mucous membrane.

Structure of nerves.   Nerves, whether in main trunks or branches, are bundles, consisting of nerve fibres in variable number; these are grouped together in primitive and secondary fasciculi. The primitive fasciculi consist of a few fibres surrounded by delicate connecting tissue, faintly fibrillated and studded with plasmatic cells; this is called the *neurilemma*, from its analogy with the *myolemma* of striated muscular fibre (Robin*). Secondary fasciculi are made up of primary bundles, which are held together by a much denser membrane consisting of ordinary connecting tissue.

In large nervous trunks, fibres of every size are found; but those of largest size are most abundant in the anterior roots of the spinal nerves, and in nerves of motion, whilst fine fibres are found in greater numbers in their posterior roots, in nerves of sensation, and in branches of the great sympathetic. These latter also contain a certain proportion of flat, pale, smooth, or faintly striated fibres, provided with well marked oval nuclei, and known as the fibres of Remak (Pl. XII. fig. II.). It is not as yet fully determined whether these are really nerve fibres, as Remak asserts, or merely a peculiar form of connecting tissue, as Kölliker and some others are disposed to think. We hold to the latter opinion, and regard the fibres of Remak as prolongations of the nucleated connecting tissue which forms the stroma of the ner-

---

* Charles Robin, Prof. agrégé d'histoire naturelle medicale à la Faculté de Medecine de Paris, Prof. d'Anatomie générale, etc., etc.—(*Ed.*)

vous ganglia (Pl. XIII. fig. III.). The nerves are not
very rich in blood-vessels; they form a net-work of
large meshes, the terminal branches from which ramify in the neurilemma; they do not come directly
in contact with the naked primitive fibre (Kölliker).

In addition to the nerve cells, which constitute Ganglions.
their most important element, nervous ganglia consist of an intermingling of very delicate connective
fibres with oval nuclei, and a large number of blood-
vessels (Pl. XIII. fig. III.). The connective element
is the same which has been already mentioned as
forming the thick nucleated external or second envelope of certain nerve cells, and the sympathetic nerve
fibres of Remak are also derived from it, for their
constituent elements are identical in form, and be-
have similarly under the influence of chemical
reagents.

It is generally conceded that nervous ganglia contain all of the different forms of nerve cells. In those
of the spinal nerves, cells with two branches (bicau-
date) are most numerous, and of these there are two
kinds. The one have their prolongations given off
from diametrically opposite points of their circumference, one being continuous with a nerve fibre from
the spinal marrow, and the other with a peripheral
nerve (Pl. XII. fig. III.). The other kind of bi-cau-
date cells give off both of their prolongations in the
same direction, and always towards the surface.
Finally, the cells with but one prolongation (unipolar) are always continuous with a peripheral fibre.
In the ganglia of the sympathetic system most of the
multipolar nerve cells seem to send off the same

number of processes, according to Leydig, as there are nerves connected with the ganglion (Pl. XII. fig. IV.). The study of the relations which exist between nerve cells and nerve fibres is one of difficulty; very thin sections must be made of fresh specimens of ganglia, and treated with dilute acetic acid, or caustic potassa, or still better, perhaps, by a solution of carmine in liquid ammonia, as recommended by Jacubowitsch.* Similar sections may be made of ganglia hardened in chromic acid; and very minute specimens of ganglia are sometimes immersed in potash and moderately compressed between two slips of glass.

Nervous centres. The arrangement of the elements of nerve-tissue, throughout the cerebro-spinal centres, is far from being clearly made out. The spinal cord consists of a central mass of grey matter, surrounded on all sides by white matter. The latter is made up almost exclusively of nerve fibres, with some blood-vessels, supported by scanty and delicate connecting tissue. By the aid of transverse and longitudinal sections, we can see that some of the nerve fibres run parallel with, and others perpendicular to, the axis of the cord. The former constitute the greater proportion of the white substance, and are found everywhere; whilst the latter are only to be seen where the roots of the spinal nerves become continuous with the substance of the cord. It is to be noted also that the

Spinal marrow.

White substance.

---

* Jacubowitsch and Owsjannikow are connected with one of the Russian Universities, and published their researches on the nerves in the Bull. de l'Acad. de Petersbourg, classe, phys.—Mathem., tom. xiv. No. 323, page 173.—(Ed.)

nerve fibres of the spinal cord are of the smaller varieties.

The grey matter of the cord forms a quadrangular Grey matter. prism with hollowed sides, or rather, a fluted column. In its centre is a canal, sometimes closed in the adult, and of greater diameter at either extremity of the cord, than in its middle. Its walls are formed by a layer of connecting tissue of varied thickness, rich in plasmatic cells, and lined internally by a stratum of cylindrical ciliated epithelium. Virchow, Kölliker, and Leydig* have particularly directed the attention of microscopists to this cylinder of connecting tissue which is thus inclosed in the grey matter of the cord, as the plasmatic cells which it contains are often the origin of morbid formations which occasionally involve its substance.

The grey matter itself is made up of blood-vessels, Relation be-tween nerve cells fine nerve fibres, and caudate nerve cells, the largest of and fibres, in the cord. which are found towards the extremities of the anterior horns. In man the connexions of the caudate cells have not as yet been clearly demonstrated. This is not true, however, of some of the lower animals; thus, in accordance with the researches of Owsjannikow, in fishes, each cell has five prolongations, which are disposed as follows: The most internal process, after it leaves the cell, enters the white commissure of the cord, and traversing it, reaches the white substance of the opposite side, where it loses itself in another similar cell (Pl. XII. fig. V. 7); this is the branch which establishes a route of connexion

* Professor of Comparative Anatomy in the University of Würzburg. —(Ed.)

between the two lateral masses of nerve cells of either side. The anterior process runs into the anterior, and the posterior, into the posterior root, of each spinal nerve; and finally, the processes given off above and below become continuous with the longitudinal fibres of the cord (Pl. XII. fig. V.). It remains now to be proved, whether or no a similar disposition exists in the mammalia and in man.

In the encephalon.

The facts ascertained in relation to the minute structure of the nervous tissue of the encephalon, are still more incomplete. Here, as in the spinal marrow, there is a white substance possessing very little vascularity, and composed entirely of nerve fibres, and a grey matter consisting of vesicular elements mingled with nerve fibres, and very numerous blood-capillaries. Besides the multipolar or caudate cells, whose processes anastomose with each other, and become continuous with nerve fibres, there exist large numbers of nuclei and spherical cells—especially in those portions of the grey matter of the brain and cerebellum which lie nearest the surface (Pl. XIII. fig. II.). What is the nature and uses of these nuclei? Are they germs of future nerve cells, or are they only the analogues of the oval nuclei which we found mingled with the connective fibres of the ganglia? These are questions which remain to be answered. As for the arrangement of the nerve fibres of the encephalon, it is probable that they run from one mass of grey matter to another, serving as commissures—but this is not positively established.*

* Refer on this subject to the admirable researches "On the Minute Structure and Functions of the Spinal Cord," by J. L. C. Shroeder van der

The ventricles of the brain and the aqueduct of Epithelium of the ventricles. Sylvius, which are properly to be regarded as the continuation, in the cranium, of the central canal of the spinal cord, are, like it, lined by a layer of ciliated epithelium. This epithelium, in some cases, is truly ciliated only in the fourth ventricle ; at least this is to be inferred from the researches of Leydig upon the brain of a criminal. (*Gazette hebdomadaire*, 1854,. p. 637). Beneath the epithelial layer there are some fibrillæ of connecting tissue forming an exceedingly delicate lamina, in which amyloid corpuscles are found. (Virchow.)

The membranes by which the cerebro-spinal ner- Membranes of brain and spinal cord. vous centres are enveloped, are composed of connective and elastic fibres, woven together in layers of different degrees of density. They have but few blood-vessels of their own, and still fewer nerves ; as to lymphatics, their existence has not even been demonstrated. The external surface of the visceral layer of the arachnoid is invested with pavement epithelium, which passes from it upon the free surface of the dura mater, where it alone forms the parietal layer of this serous membrane.

The corpuscles of Pacchioni, found in the dura Corpuscles of Pacchioni. mater along the course of its great longitudinal sinus, consist of very dense connecting tissue, enclosing sometimes, in its meshes, amyloid corpuscles and calcareous concretions.*

Kolk, Professor in the University of Utrecht, published by the Royal Academy of Sciences at Amsterdam, and translated and republished by the New Sydenham Society, London, 1859.—(*Ed.*)

* The *pineal gland* contains pale rounded cells without any processes,

Development. Nerve fibres are developed, like the other tissues, from embryonic cells. These cells unite by their extremities to form tubes, whilst their contents are being transformed into nervous marrow and axis-cylinder. We learn from the investigations of Kölliker, made upon the tail of the tadpole, that the peripheral extremities of nerve fibres take their origin independently of branching or star-shaped cells, and thus we have a clue to the mode of formation of their terminal divisions and loops. As to the development of nerve cells, it is effected by a simple change of shape and volume of primordial cells.

Reproduction. Nerve fibres, when divided, are reproduced, but the exact mode in which the process is accomplished is not well ascertained. It is asserted by some that the peripheral extremity of the injured fibre disappears, and is replaced by a newly formed fibre; others again suppose that the medulla alone is the seat of change, and that the regeneration of the fibre is effected by the formation of a new medulla in the original tube. Farther research is obviously required for the elucidation of this point of histogenesis.

and but a few nerve fibres $\frac{1}{1000}$th to $\frac{2}{1000}$ths of a line in diameter; and, for the most part, a large quantity of sandy particles.

The *pituitary body* contains, in its anterior reddish lobe, no nervous elements, but rather, according to Ecker, the elements of a vascular gland. The posterior smaller lobe consists of a finely granular substance, with nuclei and blood-vessels, and possesses, also, fine varicose nerve-tubes, which, like the vessels, descend to it from the infundibulum. Kölliker, last edition, p. 233.—(*Ed.*)

CHAPTER VI

Vessels.—Arteries.—Veins.—Capillaries, and Lymphatics.

VESSELS are of two species: blood-vessels and lym- Classification. phatics. The former are sub-divided into arteries, capillaries, and veins; the latter into lymphatics, properly so called, and lacteals. The structure of arteries, veins, the larger lymphatics, and lacteals, is almost identical; the same is true of the capillaries and smaller lymphatic vessels.

Each one of the tissues which we have studied thus far has some special or characteristic element; but this is not the case with the organs now under examination. No anatomical element of determinate form belongs exclusively to them; but that which serves to distinguish them from other tissues, is the peculiar arrangement of the several parts of which they are-composed.

Both recent and dried specimens are required for Preparations. the examination of their structure. From portions of dried vessels, very delicate and thin slices are cut by a razor, and soaked in water before being placed under the microscope. These sections answer for the examination of the outer and middle coats, and for the deeper portions of the internal coat of large vessels; but when we wish to study the epithelial layer, and vessels of very small size, such as capillaries, recent specimens are required.

Arteries.

SECT. I. ARTERIES.—When a section, either transverse or longitudinal, including the whole thickness of the wall of an artery, is placed under the microscope, we recognise three distinct strata overlying each other, which correspond to the three coats of which the vessel is composed (Pl. XIV. fig. IV.). The first, which is the thinnest, and uniformly dark throughout its whole thickness, represents the internal coat (Fig. IV. 1). The second stratum, which is transparent, and much thicker than the first, is the middle coat (Fig. IV. 2). And, finally, the third stratum, at least as thick as the second, and darker in its deeper than in its more superficial portion, corresponds to the external coat. By employing a magnifying power of 300 to 400 diameters it is easy to make out the nature, as well as the arrangement of the elements which constitute each of these coats. The following is a summary of their microscopical

Internal coat.

analysis: the internal tunic is limited, on its free surface, by a layer of simple epithelium, which, examined *in situ*, seems to consist of oval nuclei, imbedded in a structureless substance; the walls of its cells are not distinguishable in consequence of their extreme paleness (Pl. XV. fig. I.). But, by teasing out this membrane by the aid of needles, some cells may be detached, which are recognizable as fusiform in shape, with a very prominent bulge opposite the situation of their nuclei (Pl. XV. fig. II.), and which, in this respect, resemble certain cells of the spleen. Beneath this epithelial layer, which is in contact with the blood contained in the vessel, there is another lamella known by the name of *fenestrated* membrane. It is amorphous, elastic, traversed by numerous openings which vary both in size and shape, and contains elastic fibres which are disposed at right angles to the axis of the vessel (Pl. XV. fig. III. 1, 2, 3, 4). The deepest layer of the internal coat is composed of fine elastic fibres, which run in the direction of the length of the vessel. This layer is the thickest of the three laminæ which constitute the internal coat, especially in the larger arterial trunks (Pl. XIV. fig. V. 1; Pl. XV. fig. III. 5; fig. IV. 1). The semilunar valves and the endocardium are formed by the internal coat.

The middle coat is made up of elastic fibres, and Middle coat. those of non-striated muscle. The first are distributed uniformly throughout the thickness of the layer, but seem to run in no determinate direction; this is readily recognised by comparing transverse and longitudinal sections under the microscope (Pl. XIV. fig. V. 2; fig. VI. 1; Pl. XV. fig. IV. 6). The network formed by these fibres is closer in its meshes in proportion to the calibre of the artery to which it belongs, and in these meshes the muscular fibres are contained. To bring the latter into view it is well to treat the specimen with dilute acetic acid. In transverse sections it is difficult to distinguish the outlines of these fibres on account of their extreme paleness; but their nuclei, club shaped, and arranged perpendicularly to the axis of the vessel, are easily recognised (Pl. XIV. fig. V. 3). In longitudinal sections, the outlines of the smooth muscular fibres are

5

much better seen; they form polygons of variable regularity of outline, in which the central nucleus is pretty uniformly apparent (Pl. XV. fig. IV. 4). It is to be remarked that these muscular fibres are distributed with perfect regularity throughout the whole thickness of the middle coat.

External coat. The external coat is formed by a close interlacement of connective and elastic fibres, resembling felt. The farther from its outer surface the greater the amount of elastic fibres, and they generally run parallel with the axis of the vessel (Pl. XV. fig. V. 1, 2; fig. VI. 1, 2).

In reviewing the structure of the walls of arteries it is obvious that elastic fibre forms the frame-work of all their coats; but also, that in each separate coat it is associated with another distinct and characteristic element: in the internal, this is epithelium; in the middle coat, muscular,—and in the external, connective, fibre. Just in proportion as we approach the smaller terminal arterial branches, the elastic fibres tend to disappear, especially in the middle coat, which finally becomes entirely muscular (Pl. XIV. fig. VII).

In the last and smallest branches of the arterial tree which we can examine, those, for example, which measure from $\frac{1}{40}$th to $\frac{1}{30}$th of a line in diameter, we still recognise the three coats, but each one of them is constituted by only a single lamella of tissue, comprising but a solitary anatomical element. Thus, the outer coat consists of a very thin layer of connective fibres mingled with some plasmatic cells (Pl. XV. fig. VII. 1). The middle coat shows very short fusiform fibres of non-striated muscle, indicating that, in these little arterioles, this tissue remains permanently in an imperfectly developed condition (Pl. VII. 2, 3). As for the inner coat, it is reduced to a mere layer of epithelial cells (fig. III. 4).

SECT. II. VEINS.—The structure of the veins follows the same general plan as that of the arteries. Like them, they have three coats. The epithelial layer of the internal coat is identical in every respect with that of the arteries. In almost all the specimens which I have examined, a fenestrated membrane has been present, showing numerous openings, surrounded by an interlacement of large elastic fibres (Pl. XVI. fig. IV. 1, 2). *Veins.* *Internal coat.*

Beneath this is a third layer of fine elastic fibres, forming a somewhat looser web than the corresponding layer of the inner coat of an artery, and penetrating, by its deepest fibres, the surface of the middle coat—so that the dividing line between these two coats is not so distinct and clear as in the walls of an artery (Pl. XVI. fig. II. 2).

The middle tunic presents an intermixture of elastic and muscular fibres, but the latter are not uniformly distributed (fig. II. 5, 6). Their direction is generally transverse, but nevertheless, near its outer surface, there are some which run parallel with the axis of the vessel (fig. II. 7, 8). May not this unequal distribution of muscular fibre in the walls of veins account for the relative weakness of certain portions of them, and thus explain their tendency to become varicose ? *Middle coat.*

The external coat is similar in every particular to *External coat.*

that of the arteries; but it is to be noticed that in certain veins, principally in those belonging to the portal system, muscular fibres have been found in its deeper portion, longitudinal in their direction. The presence of these muscular fibres, and the direction of their course, explains the reason why these veins diminish in length under the influence of the stimulus of galvanism.

Veins of the smallest size. In veins of the smallest size the three tunics are still distinguishable; the innermost is a simple epithelial layer; sometimes, however, there is a fenestrated layer outside of it, presenting exceedingly delicate meshes (Pl. XVI. fig. V. 6). The remaining tunics resemble exactly those of arteries of similar calibre (fig. V. 1, 3).

Valves. The valves of veins are formed by their internal coat. A lamina of pavement-epithelium constitutes their surface (Pl. XVI. fig. III. 1). More deeply, we encounter wavy and parallel fasciculi of connective fibres, and a web of delicate elastic fibres intermingled with plasmatic cells. The latter are rendered visible by the addition of dilute acetic acid, which dissolves the connective fibres.

Vasa vasorum. The *vasa vasorum* are arterioles and veinules. According to Kölliker, they are to be found on vessels even of the smallest calibre (of the diameter of one half a line and less); they are distributed mainly to the outer coat; in the middle coat there are a few, but in the inner coat I have never seen them. The nerves which supply the walls of vessels are few in Nerves. number, and those which are encountered are for the most part destined to the organs supplied by the vessels which they accompany, rather than for the innervation of the vessels themselves. They seem to terminate by free ends, and it is uncertain whether or not they reach the internal coat.

SECT. III. CAPILLARIES.—The capillary vessels, Capillaries. which are the media of communication between the arteries and veins, are exceedingly simple in their structure. They are tubules of structureless substance, studded with oval nuclei. The larger the size of the capillaries, the thicker are their walls, and the greater the number of nuclei (Pl. XV. fig. VIII; Pl. XVI. fig. I. 1). The smallest of them have such exceedingly thin walls that they are portrayed by only a single line. The transition from arteries and veins to capillaries takes place insensibly, and by the successive disappearance of the several organized elements which constitute the three tunics of a vessel.

SECT. IV. LYMPHATIC VESSELS.—After what has Lymphatics. been said concerning the histology of arteries and veins, a few words only will be required for the description of the structure of lymphatics.

Their internal membrane consists of a simple layer of epithelium supported by a web of elastic fibres of extreme delicacy ; sometimes it appears to be reduced to epithelium alone.

Their middle coat is composed almost exclusively of muscular fibres arranged transversely ; elastic fibres are very few in number (Pl. XVII. fig. II. 2, 3). Finally, the external coat differs from that. of arteries and veins, by containing a large amount of longitudinal muscular fibre in its deepest portion (fig. II. 5 ; fig. III. 5). Their valves contain also some muscular fibres, and otherwise resemble those of the veins (Pl. XVII. fig. IV. 3).

The lymphatics, then, are seen to present the same general plan of structure as arteries and veins, with this single feature of difference, that they are richer in muscular fibre. The lymphatic capillaries, like those which carry blood, consist of tubes of amorphous substance, with oval nuclei set in their walls. Their characteristic peculiarity consists in the filiform prolongations which they give off at intervals along their course, and at their terminal extremities (Kölliker). As for the true seat of origin of the lymphatics (that of the lacteals being already understood), it is, according to M. Küss,* immediately beneath the several epithelial membranes—with the functions of which those of the lymphatics seem to be closely connected.

Lymphatic glands.

To the system of lymphatic vessels are attached the little gangliform organs known as lymphatic glands. These glands possess a fibrous envelope, and consist, internally, of a cortical and a medullary substance. The cortical substance, which in section presents a granular aspect, comprises the superficial portion of the parenchyma of the gland. It is a sort of extremely delicate cavernous body, whose trabeculæ, consisting of imperfectly developed connecting tissue, serve to support the blood-vessels and lymphatic trunks which enter it. Its cavities communicate in every direction with each other, with those of the

* Professor of Pathological Anatomy in the Faculty of Medicine of Strasbourg, France.—(Ed.)

central medullary portion, and with the terminal branches of the afferent lymphatics; they contain an alkaline liquid, and organized corpuscles, amongst which we can distinguish little cells, averaging $\frac{1}{115}$th of a line in diameter, and spherical granular nuclei of from $\frac{1}{415}$th to $\frac{1}{315}$th of a line. These elements are identical with those which exist in lymph and chyle.

The medullary substance, striated in appearance, is enveloped on all sides of the cortical substance, except at those points where the afferent vessels enter the gland. It is made up, mainly, of the terminal radicles of the afferent vessels, which, taking their origin in the deeper cavities of the cortical substance, anastomose with each other and form a network, whose branches, becoming gradually larger in size, and fewer in number, finally terminate in one or two efferent lymphatic vessels, which make their exit at the hilus of the gland.

The arteries, most of which terminate in the cortical substance, either arrive there directly, or after traversing the medullary portion of the organ, to which they give off a few branches. In the trabeculæ of the cortical substance these vessels form an intricate capillary plexus, which is directly in contact with the cells of the gland.

The veins, fewer in number and greater in volume than the arteries, accompany them in their course. The nerves are not numerous; they enter the gland with its vessels, and their mode of termination is unknown.

The structure of a lymphatic gland may be summed up as follows: the essential or secreting portion of the gland is represented by a large cavity (cortical substance) filled with globules, which are imbedded in a vascular network, from which they extract material for elaboration. On one side, this cavity is in communication with the afferent lymphatic vessels, and on the other, with the efferent vessels, into which latter it pours the organized products which afterwards become white, and perhaps, also, red globules of the blood.

Development of vessels.

Arteries, veins, and lymphatic vessels are alike developed from embryonic cells. They are at first recognisable in the shape of rows, or columns, of cells. Of the cells which correspond to the axis of the column, one portion liquifies, and the remainder becomes transformed into blood globules. Those which lie on either side of the axis undergo various changes, and finally form the three coats which constitute the walls of the vessel.

The development of capillary vessels is accomplished in the following manner: cells of oval shape become united to each other end to end, and then the partitions which separate them are absorbed and disappear; so that each series of cells is thus transformed into a minute canal, in the structureless walls of which nuclei are to be recognised, at regular intervals, as in the fully formed capillaries of the adult. Anastomoses are effected by means of minute prolongations, like canaliculi, which take their origin from the walls of the vessel, and, pushing out in different directions, unite finally with each other.

Kölliker has also demonstrated that branching plasmatic cells not unfrequently form a connexion, by their prolongations, with the walls of capillaries already in existence, and thus contribute to the formation of the capillary plexus, or network. Their prolongations, increasing in diameter, ultimately become true capillary vessels, and the body of the cell itself corresponds to the point of confluence of several vascular canals. The same author asserts that many large vessels are formed out of capillaries, by the transformation of the cells which surround them into their several tunics.

BLOOD AND LYMPH.—In a histological point of view the composition of blood and lymph is exceedingly simple. The organized elements of the blood are of two species, red and white globules. The red globules are bi-concave discs measuring, on an average, $\frac{1}{335}$th to $\frac{1}{366}$th of a line in breadth, and $\frac{1}{1366}$th to $\frac{1}{1366}$th of a line in thickness. Their external envelope is so exceedingly delicate that it seems to be continuous with their contents. They consist internally of an amorphous substance of some density, very elastic, and colored yellowish red by hæmatine. When exposed to the air these globules become rapidly altered in form, and as a common rule, show wrinkles, or indentations, upon their surfaces (Pl. I. fig. I. 3). The red globules, according to Schmidt, constitute about one half of the whole mass of the blood.

The white globules of the blood differ from the preceding, both in size and shape. They are spherical corpuscles, with rough or tuberculated surfaces, and average in diameter from $\frac{1}{366}$th to $\frac{1}{366}$th of a line (Pl. I. fig. I. 4). Their contents, granular and transparent, include, sometimes, a nucleus so large as to

Blood and lymph,

White globules.

almost fill the globule, but most frequently it
represented by several vesicles of brilliant appe
ance, which are rendered still more apparent by t
addition of acetic acid. Up to the present time d
tinctive features between these globules, and those
pus, have been sought for in vain. The relative p
portion, in the blood, of white to red globules, acco
ing to Moleschott,* is 1 to 357. In some cases
leucocythemia, it is increased to one-third, or even
two-fifths of the globular element of the blood (V
chow).

Plasma. The liquid portion, or plasma of the blood, coa
lates after escaping from the blood-vessels. One p
tion remains liquid, and the other assumes the fo
of a consistent and elastic mass—the clot. In t
serum, or liquid portion, a few red and white globu
are to be seen floating in the colorless fluid. T
clot, which entangles, and consequently includes, t
great mass of the globular elements of the blood,
composed, in addition to them, of a substance of
very finely granular or fibrillated appearance, but i
organized (fibrine).

Lymph globules. The solid elements of lymph are globules, wh
resemble exactly the globules contained in the c
tical portion of the lymphatic glands, and the wh
globules of the blood. These corpuscles, which a
rage in diameter from $\frac{1}{214}$th to $\frac{1}{185}$th of a line, i
found in considerable quantity in the vessels wh
emerge from the lymphatic glands, whilst in the a
rent lymphatics, or those which enter the gland a

* Professor of Physiology in the University of Zurich.—(*Ed.*)

form its capillary network, there are very few. There is found also, in the lacteals, a variable quantity of spherical granules of an oily nature, which are derived from the food, and which are very early to be seen in the true lymphatics.

Lymph, also, coagulates when it escapes from its containing vessel, forming a clot, identical in its composition to that of the blood—including, of course, no red globules. Kölliker asserts that lymph never contains red globules, and that those occasionally found in it got there in consequence of the rupture of a blood-vessel.

The formation of the red blood corpuscles in the embryo is effected, as we have already stated, by transformation of the primordial or embryonic cells which occupy the centre of the blood-vessels whilst in process of development. These cells are not at first distinguishable from the embryonic cells by which they are surrounded, but they soon become infiltrated with hæmatine, flatten out into discs, and their nuclei tend to disappear. They multiply by the process of cleavage. In the adult the increase in number of red corpuscles seems to take place at the expense of the lymph globules, which become disc-like in shape, and charged with hæmatine, whilst at the same time their nuclei are absorbed. Kölliker, relying upon what he has observed in the lower animals, asserts that this is the mode of development of the red corpuscles of the blood in man.

Formation of red globules.

CHAPTER VII.

Glands.

Definition. GLANDS are organs which present a great variety in their size and shape; they consist essentially of enclosed cavities, lined or filled with cells, and opening upon the surface of the skin, or of a mucous membrane, either directly, or by means of special canals known as their excretory ducts.

Certain organs composed of one or more cavities, closed on all sides, and filled by cells or globules, are called blood-glands, and duct-less follicles.

Division. The parenchyma of glands (the essential or secreting portion of the organ) is made up either of tubes, or of partially closed vesicles, grouped together in parcels, like clusters of fruit, and opening into a common canal or outlet. Hence we speak of two sorts of glands: those composed of clusters of vesicles (racemose), and tubular glands.

General structure. Generally glands are invested externally by an envelope of connecting tissue, varying in density. From the deep surface of this external envelope, processes or trabeculæ are given off, which, traversing the interior of the organ in different directions, divide up the glandular parenchyma into segments (lobes, lobules); they also support the vessels and nerves of the organ. The vesicles and secreting tubes are formed by a membrane of their own (basement membrane), which is usually very delicate and structure-less; sometimes it is more dense, being strengthened externally by a layer of fibrillated tissue (as in the testicles, lungs, etc.). Its internal surface is covered by a simple layer of epithelial cells, generally many-sided; or, these are arranged in strata, so as to fill completely the cavities of the secreting vesicles, or tubules. Upon the external surface of their basement membrane blood-vessels ramify, so as to form a capillary web, or network, with meshes of varying size in different glands.

Nerves accompany the blood-vessels, but are rela- Nerves. tively less numerous; their mode of termination is not yet fairly known, but it is probably by means of free extremities.

The excretory canals, or ducts, of glands, have an Ducts. external coat which is usually formed by the inter-lacement of connective and elastic fibres, together with fibres of unstriped muscle; internally they are lined by an epithelial layer, the cells of which differ from those of the parenchyma of the gland. The ducts of some glands contain minute clusters of vesi-cles in the thickness of their walls (liver, pancreas, lung).

SECT. I. GLANDS CONSISTING OF CLUSTERS OF VESI- Glands formed by clusters of vesi-cles. CLES.—The glands of this variety scarcely differ from each other in structure, except in the minute details which belong to the disposition and arrangement of their epithelial element. Therefore, in order to escape the useless repetitions which a separate de-scription of each gland would of necessity involve, we shall study the structure of certain individual glands, which will then serve as types, around which those of similar structure will naturally group themselves. And let us commence by examining those of simplest structure, *e. g.* the salivary glands.

Salivary glands, *Salivary Glands.*—On placing a very thin section of the sub-lingual gland under the microscope, we see that its terminal *cul-de-sacs*, or coecal pouches, are incomplete vesicles, the walls of which are formed by two layers. The external layer is basement membrane, structureless and exceedingly thin, being only $\frac{1}{18000}$th of a line in thickness. Its inner lining consists of a layer of many-sided cells, with extremely pale outlines, and averaging in diameter $\frac{1}{700}$th of a line. Their nuclei are much more distinct, and so large as to almost fill the cells (Pl. XVIII. fig. I. 1). Three or four of these vesicles, connected together closely in a group, have an outlet in common, and constitute thus a microscopic lobule. Several of these minute excretory canals, each with its corresponding lobule, meet together in a canal of somewhat larger size, and by their union form a lobule large enough to be seen by the naked eye (Pl. XVII. fig. V). Finally, the aggregation of a number of lobules, with a canal of proportionate size, results in the formation of lobes, and these, in their turn, uniting in a common excretory duct, make up the gland. The excretory duct is composed externally of a coat of connecting tissue, and internally of a layer of cylindrical epithelium. The gland is enveloped by a membranous expansion of connecting tissue, from the internal surface of which laminated processes are sent into the substance of the organ, where they invest the exte-

rior of its lobules and lobes, and convey to them, at
the same time, their blood-vessels and nerves.

The vessels terminate in a rich web of capillaries Vessels and
nerves.
which is spread out upon the external surface of the
vesicles. The nerves ramify with the larger vascular
trunks of the gland, but do not appear to reach their
terminal secreting pouches. We know little or nothing of the sources of distribution of the lymphatics.

To this first type are assimilated the salivary glands
and mucous follicles which belong to the cavities of
the mouth, pharynx, and œsophagus, the glands of
Brunner in the duodenum, the lacrymal glands, the
mucous follicles of the conjunctiva, vagina, and vulva,
the glands of Bartholinus and Cowper, and, finally,
those clusters of vesicles which have been mentioned
as imbedded in the walls of the ducts of the liver,
pancreas, and lung.

The Lungs.—The bronchial tubes, as is well known, Lungs.
form a tree, whose principal branches are given off
from their trunks at an acute angle, whilst their terminal ramifications are detached from the penultimate
branches at right angles. On examining the walls of
these little terminal ramifications upon their internal
surface, they are seen to be full of minute openings,
which lead into cavities (primary lobules), measuring
about $\frac{1}{80}$th of an inch in mean diameter, and which
present a somewhat complicated structure. Each of Primary lobules.
these primary lobules resembles one of the lungs of
the frog; on its inner surface we recognise large
depressions, or partially formed vesicles (Pl. XIX.
fig. I. 1), each divided into three or four secondary
vesicles (fig. I. 2), which open by large orifices into

the common cavity.   But this latter, instead of being
a large empty space in the centre of the lobule, such
as we see in the frog's lung, forms a sort of *corpus
cavernosum*, or cavernous sinus, by the anastomosis of
numerous *trabeculæ* which traverse its interior, and
which are given off from the walls of the vesicles.

· I am of opinion, however, that all of the lobules
of the lung are not constructed exactly after this
model; some of them seem to be mere *diverticula*
from the bronchial walls, presenting vesicular depressions upon their internal surfaces, but without the
trabeculæ, resembling, in short, more perfectly, the
lung of the frog.   We can often detect lateral openings near the summit of a lobule, by means of which
it communicates with neighboring lobules; but the
number of these orifices is limited.*

---

* In a monograph published within the present year (*The Anatomy of
the Human Lung, an Essay for which was awarded the Fothergillian
gold medal of the Medical Society of London, by* A. T. HOUGHTON WA-
TERS, *Lecturer on Anatomy, &c., &c., Liverpool.* Lond., 1860), containing the results of a considerable amount of original research, the true
respiratory structure of the lung is somewhat differently described.   The
trabeculæ, forming by their anastomoses a species of *corpus cavernosum*
in the common cavity of the primary lobules, are not recognised.   The
following quotations, slightly condensed from the author's own language,
give the result of his investigations.   In regard to preparations, he says
(p. 168): "The plan I have adopted, and which I believe affords the best
means for investigating the lung tissue, consists in the injection of a
colored solution of gelatine into the blood-vessels, inflation of the air-
tubes, and gradual desiccation.   In my first attempt I inflated the air-
tubes before injection of the blood-vessels, but I afterwards injected
before inflation.   The colors I have used have been red, yellow, and
blue.   The red, a finely powdered vermilion; the yellow, a chromate of
lead, formed, at the time, by the decomposition of acetate of lead by
bichromate of potash; the blue, a Prussian blue, formed by the decomposition of ferro-cyanide of potassium and sesquichloride of iron.   When

The *primary* lobules, thus described, grouped together without any intermediate substance, constitute *secondary* lobules. These latter have the shape of pyramids, whose bases are directed towards the

a piece of lung is well injected, the walls of the *air-sacs* become almost entirely opaque;· their outline may be distinctly seen when they are divided; and, the vessels in their walls being filled with the dried gelatine and coloring matter, dissections under the microscope can be carried on with great facility, without which I believe it is impossible to form a definite notion of the anatomy of these parts." With respect to injection of the air-tubes, he remarks that, " when gelatine is used as a vehicle for injecting the blood-vessels with some opaque matter, it usually happens that the coloring matter is left in the vessels, and a portion of the transparent gelatine exudes into the air-tubes; and this has appeared to me as good a way as any, of effecting an injection of this kind. Where such a preparation is dried, and then soaked in spirit and water, it swells, and assumes much the shape it has in its normal condition, and much information may be derived from an examination of it. When a piece of lung, in which the blood-vessels are injected, has injected into its air-tubes a mixture of turpentine and wax, and is left to dry, and then a slice of it moistened with Canada balsam, as suggested in Adriani's thesis (*Arius Adriani, Dissertatio anatomica inauguralis de subtiliori Pulmonum structurâ*, 1847, p. 41), the substance filling the air-tubes becomes transparent, and the outline of the *air-sacs* and ultimate bronchial tubes is well seen, and a very correct notion of their shape can be formed." . . . . "If we follow out a bronchial tube on a lung thus injected, inflated, and dried, and trace it to its termination in the ultimate air-tubes or cavities, by carefully removing the portions of lung which are upon it, and then the other half of its wall, so as to lay bare its interior, we adopt, I believe, the best plan of ascertaining how the tube itself terminates, in what manner the air cavities proceed from it, and what relation they bear to it. For this purpose, we should expose a bronchial tube, from its entrance into a lobule, to its termination. We find that the bronchial tube, having entered its lobule, divides and gives off branches, and at last terminates in a dilatation, which has opening into it a number of orifices. These orifices lead to a number of canals, which have been variously designated: 'intra-lobular bronchial ramifications' (Addison, *Philosophical Transactions*, 1842); 'lobular passages' (Todd and Bowman, v. ii. p. 390); 'intercellular passages' (Rainey *on the*

surface of the lung, whilst their summits are continuous with the bronchial passages. They are separated by a delicate interstitial lamina of connecting
tissue. Their diameter often reaches two-thirds of an

minute structure of the lungs, *Med. Chir. Trans.*, vol. xxviii. 1845);
'infundibulums' (Rossignol, *Recherches sur la structure intime du Poumon
de l'homme, et des principaux mammifères*, Brussels, 1846); 'Malpighian
vesicles' (Moleschott, *de Malpighianis Pulmonum vesiculis*, Heidelberg,
1845); and 'terminal cavities' (Mandl, *Anat. Microscop.* t. ii. ch. vi.).'"
To all of these terms the author objects, as not expressing clearly the
nature of the structure they are intended to designate; and after dis-
cussing and rejecting them *seriatim*, proposes, although unwillingly, a
new term which, he believes, expresses in a "shorter and more exact
manner than any previously used, the particular character and arrangement of the portion of the lungs under consideration:" this term is
"*air-sacs.*" "The *air-sacs* are those tubes in which the bronchial ramifications end; they are situated at the surface, and throughout all parts
of the lung; they are supported externally by the pleura, and within the
lung they in part rest, by their extremities or their sides, against the
bronchial tubes and branches of the blood-vessels, and they are visible
through the transparent coats of the smaller bronchial tubes, as through
the pleura. The *air-sacs* consist of somewhat elongated cavities which
communicate with the bronchial ramification by a circular opening, usually smaller than the cavities into which it leads, and which has sometimes the appearance of a circular hole in a diaphragm, or as if it had
been punched out of a membrane which had originally closed the
entrance to the sac; when this is the case the sac dilates suddenly
beyond the orifice. The sacs are arranged in groups (usually from six to
ten composing a group, p. 144); they are placed side by side, and separated from each other by their membranous walls; their shape, when
properly inflated, or when distended by some material which has set in
the sacs, such as gelatine, or a mixture of wax and turpentine, is polygonal;
they approach very nearly to the circular form, but in consequence of
their mutual pressure, their parietes become somewhat flattened. They
increase somewhat in size as they pass from the bronchial tubes to their
fundus, the latter being usually the broadest part of the sac; but they
are often found to have an almost uniform diameter throughout. All the
sacs pass from the extremity of the bronchial tube *towards* the circumference of the lobule in which they are placed; they consequently radiate

inch, or more. In the adult, as a rule, the outline of
the bases of these lobules is marked on the surface of
the lung by a variable amount of black pigmentary
deposit.

from the tip of each terminal bronchial twig. *The sacs connected with
one bronchial termination do not communicate with those of another ;*
each set of air-sacs is therefore a little lobule, or *lobulette*, which, in fact,
represents the entire arrangement of the lung, and is a lung in miniature."
(The resemblance of a *lobulette* to the entire lung of a frog, is elsewhere
strongly emphasized.) " As the *air-sacs* pass towards the boundary of
the *lobulette*, they often bifurcate, and here and there circular orifices
exist, which lead to smaller sacs, sometimes only to a small group of air-
cells or *alveoli*, so small as scarcely to be considered a sac." The author
adopts the term *alveolus* from Rossignol as preferable to *air-cell*, or *air-
vesicle*, both of which he considers objectionable. " A pulmonary *alveolus* is that portion of an *air-sac* which exists in its wall, and is circum-
scribed by a slightly raised margin, consisting of thin membrane, and
constituting a cup-like depression. In shape it is more or less polygonal.
The alveoli are found throughout the circumference of the sacs, and at
their fundus, varying in number in each sac from eight to twenty."
" If we trace the sacs from their fundus we may say that, passing from
the periphery of the *lobulette*, and diminishing somewhat in size, they all
terminate in the dilated extremity of the bronchial tube," by the common
orifice already described. " The sacs, as they pass in this manner, often
join, two and three together, and others terminate in a single mouth."
. . . . " The tube which results from the union of two sacs has a
smaller capacity than that of the two sacs taken together, but a larger
capacity than either of them individually." . . . . " The shape of
these groups of air-sacs, or lobulettes, is more or less pyriform, the apex
being situated at the termination of the bronchial tube ; the base,
somewhat flattened, especially at the superficies of the lung, at the distal
extremity of the sacs. A most excellent way of examining the air-sacs,
and one which demonstrates most satisfactorily the manner in which those
at the surface of the lungs are arranged with reference to the bronchial
tubes, is the following : a thin slice should be cut off the surface of a
portion of lung which has been injected, inflated, and dried, and the
portion itself (not the slice) should then be placed under the dissecting
microscope. The cut orifices of the air-sacs will be observed. Very
fine bristles should then be inserted into these tubes, the largest one being
first chosen. It will be found that several of the bristles, passing into

The constitüent parts of a pulmonary vesicle, passing from without inwards, are, 1st, basement membrane; 2d, its epithelial investment. The first is
formed by an extremely delicate web of elastic fibres;

different openings, converge to a point a short distance from the surface
of the lung. It will be known that they pass to a common point, as, by
moving one of the bristles gently, it will act upon the others. Having
placed bristles in all the air-sacs which converge to this spot, the bristles
should be left in their position, and the bronchial tube should be laid
bare on its proximal side, down to its termination, *i. e.* the lung substance covering it should be removed, care being taken to stop just
before reaching the spot where it communicates with the air-sacs. The
number of bristles communicating with the spot thus exposed, will show
the number of air-sacs belonging to one group. The termination of the
bronchial tube will be seen to be somewhat expanded, and the air-sacs
will be found, many of them, to communicate with it by a circular orifice, which is smaller than the sac itself." . . . . . " When the bronchial tube has been exposed in the way I have mentioned, and the mode
of communication with the air-sacs observed, a section may be made
longitudinally through one or more of the latter; it will be then seen
that the sacs lie side by side, and that they occasionally give off smaller
sacs; the manner also in which they divide, and the mode in which
they terminate, will be observed. It will also be seen, that as each
bronchial tube approaches its termination, it has here and there throughout its circumference, small circular orifices, which are the commencement
of small canals, leading to groups of air-sacs or lobulettes; and it will
further be seen that the tube itself, at its termination, has a number of
alveoli in its walls.

"Another very excellent way of examining the terminal bronchial
tubes, and the commencement of the air-sacs, is to soak a piece of lung
that has been injected, inflated, and dried, in spirits for some time, and
when the piece is well saturated, to dissect it under the microscope. By
imbibition of the spirit the mass of lung swells, and the air tubes and sacs
remaining distended, the parts assume nearly the size and shape they have
in life. When such a piece is examined, very frequently on opening the
bronchial tubes and following them to their end, *their* alveoli may be
plainly seen, as well as the orifices leading to the air-sacs, and the band
of elastic fibres which surrounds the opening into each sac becomes
apparent."—pp. 132–147.—(*Ed.*)

these accumulate in the intervesicular partitions, and form likewise the central portions of the trabeculæ Pl. XIX. fig. II. 1). The nature of this membrane explains the great elasticity of the lungs.

The epithelial lining of the pulmonary vesicles is composed of many-sided cells with very pale outlines, measuring, in mean diámeter, $\frac{1}{810}$th of a line, and having no cilia; their nuclei are very large ($\frac{1}{810}$th of a line) and full of dark granules (Pl. XIX. fig. II. 2; fig. III.). This epithelial layer is found also upon the trabeculæ. I have reason to believe that fatty degeneration of these epithelial cells always constitutes the initial lesion of pulmonary tuberculosis.* *Epithelium.*

In tracing the epithelium from the vesicles and lobules into the bronchial tubes, the single layer of cells becomes double, and the number of strata continues to increase as the air-tubes enlarge in calibre; *Bronchial epithelial layer.*

---

* The existence of an epithelial lining to the air-cells of the lungs is one of the most recently established facts of histological science. It is denied by recent and high authorities, viz. Rainey (*Med. Chir. Trans.* vol. xxxii. 1849, p. 51; and *Brit. and For. Med. Chir. Rev.* No. xxxii. p. 491), and Todd and Bowman (*Phys. Anat.*, Lond. 1856, vol. ii. p. 391); although the latter admits the fact in his article on " Mucous Membranes" in the Cyclopædia of Anatomy. It is asserted by Carpenter (*Human Physiology*, p. 513, 4th Ed. Lond.), by Quain and Sharpey, Kölliker, Rossignol, Adriani, Schrœder van du Kolk, Schultz (*Disquisit. de structurá et texturá canalium æriferorum*, 1850, p. 10); Williams (*art. Lungs in Cyclopædia of Anatomy*, 1855); Dr. Radclyffe Hall (*Brit. and For. Med. Chir. Rev.* No. xxx. p. 481); Mandl (*Anat Micros.* vol. ii. p. 327); Milne Edwards (*Leçons sur la Physiologie, &c.*, t. ii., p. 326); Peaslee (*Human Histology, &c., &c.*, Philad. 1857, p. 579), and Waters (*op. cit.*, p. 159). According to the latter authority it is best seen in a perfectly fresh specimen of lung tissue, and by the aid of acetic acid. According to Kölliker it is "an ordinary pavement epithelium without cilia, which forms a single layer, and rests immediately on the fibrous coat" of the pulmonary vesicles.—(*Ed.*)

of these, the deeper layers of cells are many-sided, and present nothing worthy of note; but those upon the surface are conical, with their bases directed towards the axis of the canal, and furnished with vibratile cilia (Pl. I. fig. VII.). In the most minute bronchial tubes the deeper strata of epithelial cells disappear entirely—the ciliated layer alone remaining.

Cilia.

Underlying the epithelial layer we have the mucous membrane, consisting of a delicate web of intermingled connective and elastic fibres. These latter have a longitudinal direction, and occupy the outer aspect of the membrane. In bronchiæ of some size they form small whitish longitudinal fasciculi, perfectly visible to the naked eye. Outside of the mucous membrane we have a layer of unstriped muscular fibres, circular in their direction. In the trachea, and its larger subdivisions, these fibres are found only in their posterior aspect, or in the membranous portion of the canal, and their connexion with the extremities of the cartilaginous rings has been demonstrated (Kölliker). Here, also, longitudinal muscular fasciculi have been recognised, occupying the outer aspect of the circular layer.

Mucous membrane.

Finally, the external or fibrous coat of the bronchial tubes is formed by a dense interlacement of connective and elastic fibres. It contains also cartilaginous plates of variable shape, which are found only upon the anterior and lateral regions of the trachea and large bronchi, whilst they are distributed upon the whole circumference of the smaller canals. It is to be noticed that these cartilaginous lamellæ become smaller and less frequent in proportion as the

Fibrous coat.

tubes diminish in calibre, and finally, when this has reached a diameter of one-half of a line, they are no longer to be found. Tubes of this size, in fact, consist simply of an extremely delicate mucous membrane, lined externally by scattered muscular fibres, and within, by ciliated epithelium.

Clusters of gland vesicles are found imbedded in the thickness of the walls of the trachea and bronchi. Very numerous in the commencing trunks of the bronchial tree, they become less and less frequent as its branches diminish in size, and, according to Kölliker, are no longer to be seen in bronchiæ of from one to one and a half lines in diameter. These little glands, which are hardly one-fourth of a line in diameter, are to be found in the deepest portion of the mucous coat of the tube, or rather lying upon the inner surface of its fibrous coat. Their epithelial cells are polygonal, whilst those lining their ducts, which open into the bronchial tubes, are cylindrical, but not ciliated.* *Glands.*

The pleura, like the other serous membranes, are not very complicated in their structure. They are made up of a somewhat dense interlacement of fibres —connective and elastic—and, upon the free surface of the membrane thus formed, a simple layer of pavement epithelium. They receive numerous blood- *Pleura.*

---

* Waters (*op. cit.* p. 122) describes *two* sets of glands as belonging to the mucous membrane of the trachea and bronchial tubes : one consisting of simple mucous follicles, found everywhere on the surface of the membrane ; the other, larger in size and compound in character, found only in the posterior, or membranous portion of the trachea and larger bronchial ramifications. It is these latter which are supposed to furnish the very tenacious and viscid masses of sputa so characteristic of certain stages of bronchitis.—(*Ed.*)

vessels from a variety of sources (the bronchial and pulmonary arteries, intercostals and internal mammaries), which enter at their attached surfaces. Finally, nervous filaments have been traced into their substance from the great sympathetic, the *nervus vagus*, and the phrenic nerve.

Pulmonary artery. The arteries distributed to the lungs are of two sorts, bronchial and pulmonary arteries. The latter accompany the bronchial tubes to their terminal extremities, and, during their course, subdivide very frequently, some of their branches going to the smallest of the bronchiæ, whilst the rest terminate on the pulmonary vesicles. Before they finally break up into a capillary plexus, the smaller arterial branches are found in the interstices between the lobules of the lung, where they anastomose with each other in such a manner as to surround each lobule with a vascular circle or network, recalling the arrangement of the branches of the *vena portæ* around the lobules of the liver. From this arterial circle branches of the smallest size are given off in great number, which, by their inosculations, form a capillary network with exceedingly small meshes ($\frac{1}{160}$th of a line in diameter), which occupies the deep layer of the walls of the pulmonary vesicles. Of these ultimate branches of the pulmonary artery, some leave the lobules to supply the visceral layer of the pleura.

Veins. The radicles of the pulmonary veins, which take their origin from the capillary plexus, spread themselves upon the surface of the pulmonary vesicles, forming a stratum more superficial than the capillaries; then they lose themselves in the lobular interstices, where they unite with each other to form larger branches, which finish their course afterwards, either alone, or by joining company with branches of the pulmonary artery.

The bronchial arteries supply the whole bronchial tree, and also the pleuræ. Some of their terminal branches, those which supply the smallest of the bronchial canals, anastomose with branches of the pulmonary arteries and veins; others terminate by corresponding veinules* through which their blood is ultimately brought back to the *vena cava superior*. Bronchial arteries.

The lymphatics of lungs form two sets, one of which is superficial, ramifying upon the pleuræ, and the other deep, and accompanying the bronchi and large vessels. In the intervals between the lobules numerous anastomoses take place between these two sets of vessels, both of which ultimately reach the bronchial glands at the roots of the lungs. Lymphatic glands have never as yet been discovered in the parenchyma of the lungs. Lymphatics

The nerves of the lungs are derived from the pneu- Nerves.

---

* The existence of bronchial veins *within* the lungs is denied by Waters (*op. cit.*), whose original investigations entitle his opinions to respect. At p. 405, under the head of *bronchial veins*, he says, " The only vessels I have been able to discover to which this name can be applied, are some small ones situated at the *root* of each lung, at its posterior aspect. I have previously mentioned that their distribution has always appeared to me to be confined to the structures about the root of the lung, the bronchi, the bronchial glands, &c.; and *that they do not return the blood from the interior of the lung. I have never seen, either in the lungs of man, or those of other mammalia I have examined, any veins accompany the several branches of the bronchial artery along the bronchial tubes.*" Reisseissen (*de fab. Pulmonum*, Berlin, 1823) appears to hold the same view.—(*Ed.*)

mogastric and the great sympathetic. They ramify in company with the bronchi and branches of the pulmonary artery, and present, at intervals, minute enlargements composed of nerve cells ; their mode of termination is unknown.

Development. According to M. Coste* the first appearance of the lungs occurs in the shape of a small granulation, in the median line, projecting from the anterior wall of the œsophagus. This little mass is hollow within, and communicates with the œsophagus by means of a vertical slit which, eventually, by the dilatation of its walls, forms the larynx and trachea. Soon this central mass divides into two lateral portions to form the two lungs. Still later, each lateral mass divides itself up into an infinite number of vesicular vegetations, and is thus transformed into the parenchyma of the lung. Finally, the process of development is completed by the various metamorphoses of the embryonic cells by which the several histological elements which compose the tissue of the respiratory organs are formed.

According to Bischoff,† and most of the German embryologists, the lungs are first recognised in the form of solid sprouting buds, which subsequently become hollow by the melting down of their central cells ; the remaining steps of the process being identical with those already described.

Sebaceous glands. *Sebaceous Glands.*—Sebaceous glands, which are almost universally associated with the hair follicles,

---

* Embryogénie Comparée. Paris, 1837.—(*Ed.*)

† Professor of Physiology in the University of Heidelberg, Baden, Germany.—(*Ed.*) ⁚

are formed by a duct, which, when traced from its
orifice, is found, most generally, to remain undivided,
and to terminate in a solitary bulb, or cœcal pouch,
of considerable size (Pl. XVIII. fig. II.). The inter- Structure.
nal surface of this pouch is covered with depressions,
which represent the vesicles of the glands we have
already described. Sometimes a group of these depressions, becoming deeper, isolate themselves incom-
pletely from the common cavity, by a partially
formed neck, or pedicle, and thus constitute a lobule
(fig. III. 2). The wall proper of the sac, or pouch,
is structureless, and very thin, but it is strengthened
by an external layer of fibrillated tissue (Pl. XVIII.
fig. II. 5). In contact with its inner surface are one
or two layers of young cells, with finely granular and
transparent contents, and nuclei which are quite distinct (fig. II. 2). Upon this epithelial layer are
numerous other cells, differing from those just described, both in volume, and in their contents. They
increase in size, in fact, exactly in proportion as we
trace them nearer to the centre of the gland cavity;
and during their growth in size, the nucleus disappears, and their contents are transformed into oil-
globules (fig. II. 1 ; fig. IV.; fig. V. 2). Finally, near
the orifice of the duct, these cells, having attained the
maximum of their development, and being no longer
able to resist the pressure from all sides of the cavity,
the contents of which are thus constantly increasing,
rupture their cell walls, and give forth a sort of greasy
substance which, in short, is the true secretion of the
sebaceous gland (Pl. XVIII. fig. II. 4). The excretory canal generally opens into a hair follicle; but

sometimes it gives directly on the surface of the skin, and when this is the case, that portion of the duct which traverses the epidermis has no walls of its own (fig. II. 6).

Vessel and nerves. The vascular supply of the sebaceous glands presents nothing peculiar, and as to their nerves we know nothing of their distribution.

Development. Sebaceous glands are developed from the external epidermic layer of the hair bulb, or rather from the *rete mucosum* of the epidermis. A small projection buds forth, on the surface of which others, still smaller, make their appearance, and these, increasing in number, constitute the vesicles of the gland ; whilst the base of the primitive projection, becoming more elongated and constricted, forms its duct. According to Valentin, the earliest rudiments of the sebaceous follicles become perceptible during the last half of the fourth month of fœtal life.

Meibomian glands. The Meibomian follicles are an aggregation of minute sebaceous glands, all opening into a long, common, excretory duct (Pl. XXVII. fig. III.).

Mammary glands. The mammary gland, which in external appearance resembles the salivary glands, is identical, as regards its epithelium, with the sebaceous glands, at least during lactation. It still more closely approximates to the sebaceous glands by its anatomical position, and its mode of development.

Milk. The histological elements of milk consist of simple minute oil-globules, of very brilliant aspect, and dark, strongly marked outlines, floating in a transparent fluid (Pl. XVIII. fig. VI. 4). During the first few days of lactation we meet with a certain proportion of these oil-globules which, instead of being free and solitary, are aggregated together in little round masses, forming what are known as globules of *colostrum* (fig. VI. 2). When the gland is in a state of inflammation, these colostrum globules make their appearance in the milk. The good quality of the milk is known by the large number and equality in volume of its globules.

On reviewing and comparing the glands, whose Division of glands. structure we have thus far studied, reference being had solely to the character and arrangement of their epithelium, it is obvious that they can be divided naturally into two groups: 1st, glands with simple epithelium; 2d, glands with stratified epithelium. In each of these varieties the process of secretion is differently accomplished. In the first group, the plasma of the blood exudes through the epithelial layer, is modified by its cells, and passes out through the excretory duct without carrying with it any solid elements; this is the process of secretion *by simple filtration*. In the second group, in which the epithelial cells are packed in strata, the blood plasma, in passing through them, excites in them a higher grade of vital action; they increase rapidly both in size and number, and their contents undergo at the same time a specific change; but both the cells and their contents ultimately become disaggregated, melt down, as it were, and thus form the secretion—which is hence called *secretion by epithelial growth*.

The salivary glands, and the numerous family of mucous follicles, effect their secretion by the mode first described; the sebaceous glands, and the mammary gland, alone belong to the second class. We shall see hereafter that these considerations are also equally applicable to tubular glands.

Follicles of Lieberkuhn. SECTION 2d. TUBULAR GLANDS.—The most simple form of tubular glands are those of Lieberkuhn, which, as is well known, are thickly scattered throughout the whole extent of both the small and large intestine. They are simple straight tubes, one end of which opens upon the free surface of the intestinal mucous membrane, whilst the other, slightly enlarged into a bulbous sac, is imbedded in the deeper portion of this same membrane (Pl. XXVI. fig. X.). Their mean diameter varies from $\frac{1}{25}$th to $\frac{1}{15}$th of a line. Each follicle is composed of a structureless basement membrane, on the outer surface of which the blood-vessels belonging to the intestinal glands ramify, whilst within it is lined by a single layer of cylindrical epithelial cells, which are arranged very regularly around the cavity of the tube (Pl. XXVI. fig. XI. 1, 2). The cavity varies from $\frac{1}{120}$th to $\frac{1}{150}$th of a line in diameter.

Development. The development of these glands is effected at the expense of the epithelial lamina of the intestine, which is protruded like the finger of a glove in the form of a tube. They will be hereafter more fully examined in connexion with the intestinal mucous membrane.

The glands which secrete pepsin, found near the cardiac orifice of the stomach, and its mucous follicles, situated near the pylorus, are nothing more than compound follicles of Lieberkuhn. They are composed of from two to six single tubes, similar to those described above, which unite to form a common excretory duct. In the mucous glands the same variety of epithelium lines the interior of the tubes of their excretory ducts; it consists of conical cells similar to those in the glands of Lieberkuhn, measuring from $\frac{1}{310}$th to $\frac{1}{250}$th of a line, and their nuclei $\frac{1}{750}$th of a line in diameter (Pl. XXV. fig. VIII.). The epithelium of the ducts of the pepsin glands is similar to this, but the cells of their tubes are larger ($\frac{1}{155}$th of a line), and they are polygonal in shape (Pl. XXVI. fig. 1.). Sometimes they present minute projections of the basement membrane of the tubes, so that they have a mamelonated appearance externally. According to Kölliker, the epithelium of these compound glands is liable to fatty infiltration, a phenomenon which we never observe in Lieberkuhn's glands. But this is not of constant occurrence, and it is probable that the presence or absence of fatty particles in the epithelium of these glands, corresponds to periods of activity and rest in the functions of the gastric mucous membrane.

The glands of the mucous membrane of the uterus Uterine glands. have precisely the same physiognomy as the tubular gastric glands. Like them, they consist of a solitary tube, or of two, converging to a common excretory duct, and their epithelium, of a single layer of conical cells. As most of them are too long to be accommodated in the thickness of the mucous membrane, we find that their local extremities are somewhat curved, or bent upon themselves.

The glands of the neck of the uterus are not so long as those of its body. When their orifices become obliterated, their secretion, accumulating within, distends them and alters their shape, so that they become spherical, and thus form what are known as the *Ovula Nabothi*.

Sweat glands. *Sudoriparous Glands.*—These tubular glands occupy the deepest layers of the skin (Pl. XXIII. fig. I. 8). The body of the gland, or *glomerula*, is a little spheroidal mass, or ball, averaging one-fourth of a line in diameter. It consists of a tube, rolled and twisted upon itself, and terminating by a blind extremity, which, in rare cases, is bifurcated (Pl. XIX. fig. IV). Its walls are formed by an extremely thin and structureless basement membrane ($\frac{1}{1800}$th to $\frac{1}{1200}$th of a line in diameter), which is strengthened externally by some fibres of connecting tissue very rich in plasmatic cells (fig. IV̇. 4), and which may be regarded as constituting the fibrous envelope of the gland. The internal surface of the tube is lined by a single layer of pavement epithelium, whose cells measure $\frac{1}{700}$th of a line in thickness (fig. VI. 2). Finally, the glomerula is surrounded by a rich vascular network, which affords the materials for its secretion. Of the mode of innervation of these glands we are as yet ignorant. The excretory duct of a sweat gland, after leaving the glomerula, runs directly outwards through the substance of the skin, towards the bottom of one of the furrows upon its surface, and then, traversing the epidermis, terminates by an oblique opening upon its outer surface. During its course through the skin proper it is perfectly straight; but, in traversing the epidermis, it turns upon itself, forming a close spiral twist (Pl. XXIII. fig. I. 9). In the epidermis the duct has no proper coat, this being replaced by the epidermic cells which form its walls, but, in the thickness of the skin, its walls are formed by two coats: the outer, dense and fibrillated ($\frac{1}{144}$th of a line in thickness), including non-striated muscular fibres, which, according to Kölliker, are also found in the fibrous envelope of the glomerule; the inner coat thin ($\frac{1}{1500}$th of a line), unorganized, and lined by an epithelium similar to that of the gland (Pl. XIX. fig. V.; fig. VI.).

The ceruminous glands, which belong to the external ear, are identical in form with the sweat glands; they differ from them only in the nature and arrangement of their epithelial cells. Thus, instead of being spread out in a simple layer upon the internal surface of the secreting tube, their cells form a series of strata by which its cavity is completely filled (Pl. XIX. fig. VII. 2). Moreover they become infiltrated with yellow pigment, and oil-globules in abundance, characteristics in which these glomerules, as regards their secretion, resemble closely the sebaceous glands. The large sudoriparous glands of the axillæ seem to be absolutely of the same species as the ceruminous glands, for their contents are identical.

Ceruminous glands.

The sudoriparous glands are developed from the fifth to the eighth month of fœtal life. They make their first appearance in the shape of minute cellular projections from the deep surface of the epidermis, which imbed themselves in the true skin. At first they are nothing more than solid cylinders slightly bulbous at their dermal extremities. As they grow, they reach the deepest layer of the skin, but do not

Development.

pass beyond its limits; as their growth in length, however, goes on, they curve and twist upon themselves, so as to form the glomeruli which we find in the adult. Whilst these changes are taking place in the size and external shape of the sweat glands, their interior is observed to become converted into a canal, and this is probably effected by the disaggregation and melting down of the central cells. Finally, the walls of the gland tubes are perfected by a series of morphological transformations of the more superficial cells of the original cylinder.

Preparations. The specimens required for the study of the tubular glands are very readily prepared. It is simply necessary to detach, by means of scissors, very delicate little lamellæ from the mucous membrane of the intestine, cutting both parallel with, and at right angles to, its surface. For those of the skin, a razor is employed in the same manner, and dilute solutions of potassa and acetic acid are applied to the specimens in order to render the tissues more transparent.*

Kidneys, structure. *Kidneys.*—In a longitudinal section of a kidney, including its *hilus*, the parenchyma of the gland presents itself in two obviously different aspects; near the *hilus*, and towards the centre of the gland, it is striated; elsewhere it is granular in appearance. The striated portion (cones, medullary substance, pyramids of Malpighi) consists of sections of cones, of which the

---

* The spiral twist of the duct of the sweat gland, as it traverses the epidermis, is best seen by cutting thin slices from the edge of a piece of dried skin of the palm of the hand or sole of the foot, with a sharp scalpel.—(*Ed.*)

apices (*papilla*) converge towards the *hilus*, whilst their bases are directed outwards, towards the surface of the organ. The granular portion (cortical substance) forms not only the peripheral substance or cortex of the organ, surrounding the bases of the pyramids, but it dips inwards between them, reaching nearly as far as their summits. The free surfaces of the *papillæ* are pierced by numerous small orifices, $\frac{1}{70}$th of a line in diameter. Each of these orifices leads into a straight canal, which, in its course, is constantly dividing and subdividing, always into two branches, and these are given off invariably at a very acute angle. Having reached the base of the medullary cone, all of these subdivisions of the primitive tube, which thus far have pursued a perfectly straight course, immediately become exceedingly tortuous, and by their involutions and twistings, form the cortical substance of the gland, each tubule terminating at last by a bulbous extremity, which is in intimate contact with a small tuft of blood-vessels called a Malpighian body (Pl. XX. fig. I. 2, 3).

It is noticeable that the line which limits the base of each pyramid forms a series of indentations, and that the straight tubes emerging from the summit of the intervening projections, are surrounded on all sides by tortuous tubules (fig. I. 1).

In accordance with this description it is obvious that each primitive straight canal gives origin to a fasciculus of tubes (pyramid of Ferrein, lobule of the kidney) which pursue a straight course through the substance of the medullary cones (tubes of Bellini), and become tortuous in the cortical substance of the organ (tubes of Ferrein).

The straight tubes measure, on an average, $\frac{1}{12}$th of a line, their subdivisions, $\frac{1}{18}$th of a line, the tortuous tubules of the cortical substance $\frac{1}{24}$th of a line, and their bulbous extremities $\frac{1}{36}$th of a line. A perfectly structureless basement membrane, scarcely $\frac{1}{1800}$th of a line in thickness, forms the outer walls of these secreting tubules, which is hardly distinguishable except when denuded of its epithelium (Pl. XX. fig. II. 4). Its internal epithelial lining, at least ten times the thickness of the outer wall, is composed of a single layer of many-sided cells, usually pale in their outlines, and each containing a large and well-marked nucleus, which is very clearly recognisable in the midst of its transparent granular contents (fig. II. 5, 7). Fatty degeneration of these epithelial cells is the principal lesion of the urinary tubules in Bright's disease. These two membranes, alone, constitute the walls of the urinary tubules, as far as their bulbous or expanded terminations, where another element makes its appearance, which we shall proceed to examine.

Arteries. In the hilus of the kidney the renal artery divides into about a dozen branches, which, arranging themselves between the medullary cones, penetrate the cortical substance, and ultimately reach the surface of the organ. In their course, which is almost rectilinear, they give off a great many branches to the lobes, as they traverse the spaces between them, and some smaller arterioles, also, to the fibrous envelope of the gland. The ultimate branches of the renal artery, having gained the interior of the medullary pyramids, shortly after their origin, pursue a course parallel with the straight tubes of which they are composed, and continue onward through the cortical substance, towards the periphery of the gland. Whilst in the pyramids there is nothing peculiar in their distribution, but, as soon as they reach the cortical substance, they begin to give off small branches, at regular intervals (afferent vessels), in every direction, which penetrate the walls of the secreting tubuli, and occupy the interior of their expanded extremities (Pl. XX. fig. III. 1 ; fig. IV. 4, 5). Here, each afferent vessel breaks up into a certain number of branches, which, becoming exceedingly tortuous, intertwine with each other so as to form a little round ball, or tuft, which is known as the Malpighian body (glomerula of Malpighi, *corpus Malphigianum*). Malpighian bodies.

These little vascular balls completely fill the pouchlike terminal expansions of the *tubuli uriniferi*, and it can be recognised that their surfaces are entirely covered by a layer of renal epithelium. We have succeeded in demonstrating this relation between the Malpighian tufts and the epithelium of the urinary tubules on several occasions, in the kidneys of the guinea-pig, and our researches into the structure of the human kidney have led to the same result (Pl. XXVII. fig. IV. 5).

A solitary vessel, of capillary size (the efferent vessel), leaves the Malpighian tuft, traversing the wall of its containing cavity, either alone, or in company with the efferent vessel, near which it is always found ; it immediately divides into a multitude of ramifications, which anastomose with each other, and Efferent vessel.

with those of the neighboring efferent vessels. By this system of anastomosing vessels the capillary network of the kidney is constituted, and the urinary tubules are everywhere closely surrounded by it; in the cortical substance this network is very fine and close (Pl. XX. fig. III. 5), but in the medullary cones, its meshes grow longer, and the number of the vessels diminishes (fig. III. 6). They continue to diminish in number, and to increase in size, forming venous radicles, which, assuming a straight course, ultimately empty into the renal vein.

Lymphatics. The origin of lymphatics from the kidney is imperfectly made out, and the same is true of the terminations of its nerves, which enter the organ at its hilus in company with its vessels.

We may sum up the structure of the kidney in a few words, as follows : from a papillary orifice a canal takes its origin, which, leaving its sub-divisions out of view, is rectilinear in the medullary cones, becomes tortuous in the cortical substance, and ends there in a flask-like pouch, in the interior of which the Malpighian tuft is lodged. This takes its origin from one of the interlobular arterial branches by means of the afferent vessel, and itself gives off the efferent vessel, which is the source of the capillary system of the organ, by which its secreting tubules are enveloped, and which pours its blood ultimately into the renal vein. The secreting tubules have two walls : the one thin and structureless, the other much thicker, and consisting of a layer of epithelial cells which, at its expanded extremity, are found covering the entire surface of the Malpighian tuft. Nothing certain is known of its nerves and lymphatics (Pl. XX. fig. IV.).

The proper coat of the kidney consists of a dense *Fibrous coat.* interlacement of connective and elastic fibres; its internal surface is united to the cortical substance of the organ by minute vessels, and delicate trabeculæ of its own tissue; its outer surface is continuous, by means of slender fibrous processes, with the mass of adipose tissue in which the gland is imbedded; at the bottom of the hilus it becomes continuous with the *calices.*

The secreting duct of the kidney, or *ureter*, is *Ureter.* expanded above, where it forms the *pelvis* and *calices,* and terminates below at the bladder. On examining into the structure of its walls, we find, proceeding from without inwards: 1st, a fibrous tunic; 2d, a layer of smooth muscular fibres—composed externally of longitudinal and internally of circular fasciculi; 3d, a mucous membrane, with its epithelium in strata, and becoming sensibly thinner where it passes from the surface of the calices upon the papillæ. Of this epithelium the deep cells are very regularly oval, but those nearer the surface are variable both in size and shape, and resemble exactly what are called cancer cells; the epithelium of the bladder, which is continuous with the ureter, presents the same physiognomy; that of the urethra is composed of cylindrical and oval cells, of regular form.

In the unmixed urine, as it escapes from a healthy *Urine.* kidney, histologically, there is nothing to be recognised. It is a liquid in which we find no organized element, unless it be here and there a cell, accidentally detached from the walls of the ureters, bladder, or urethra, and carried along with the urine in the act of micturition.

Development.   The kidneys are developed behind the Wolffian bodies, and entirely independently of them.   They take their origin from the mucous membrane of the intestine, but the facts which we possess relative to the transformations which these organs undergo during their development, are not clearly enough established to justify their introduction into this work.

Preparations.   The most useful preparations for the study of the structure of the kidneys, are sections made with a razor of kidneys rendered solid by boiling, and similar sections of fresh organs, as well as of those previously injected with fine colors ground in oil and well rubbed up with pure oil of turpentine.

[The most recent and valuable additions to our knowledge of 'the minute structure of the kidney are due to the industry and talent of the late Dr. C. E. Isaacs. His admirable paper, entitled " *Researches into the structure and physiology of the kidney* (by C. E. Isaacs, M.D., Demonstrator of Anatomy in the University of the City of New York,") was read before the New York Academy of Medicine in March, 1856, and in its numerous and excellent illustrations the student will find great assistance in thoroughly comprehending the minute anatomy of this organ.

In view of the close and searching investigations of Dr. Isaacs into the doubtful points in the minute structure of the kidney, and his high character as a conscientious and skilful observer, I feel that no apology is necessary for stating his conclusions at length, and describing the ingenious and original methods by which he arrived at them.

He speaks of the "highly-elastic" nature of the basement membrane forming the outer wall of the urinary tubules, and adds, " the integrity of the tube being of the highest importance during life,

this membrane has been endowed with the quality of strongly resisting injurious influences, and even powerful chemical reagents. In
virtue of this last-named property, we are enabled to show clearly
the tubes of the kidney, by certain processes hereafter to be described."—*Trans. N. Y. Acad. of Med.*, Vol. I. part IX. p. 379.

In relation to the nature of the renal epithelium, it is stated
(p. 380) that "most of the epithelial cells are polygonal, although
many are oval, and others of a rounded or irregularly rounded
shape." . . . "It is extremely difficult to meet with specimens of
the kidney sufficiently healthy to exhibit the perfectly normal
epithelium." . . . "The epithelium very soon becomes changed
from decomposition, or by the action of water, which expands the
cells, and sometimes causes them to burst, when the tubes are found
to contain merely nuclei and granular matter. In examining the
epithelium, *it is therefore very important* to obtain the kidney in as
fresh a condition as possible, and instead of water, to use a solution
of albumen in water, or urine." In conclusion, the opinion is
expressed, that all the appearances presented by the renal cells
differing from the characteristics of pavement, or tesselated epithelium, are the result either of decomposition or disease. (p. 381.)

As to the presence of *cilia* upon the epithelial cells of the kidney,
it is admitted, in fishes and the amphibia. To determine the question as to its existence in the mammalia, he "resorted to the large
establishments, in this city, for killing oxen, sheep, horses, dogs,
rats, etc., and examined the kidneys immediately after the death of
the animal." . . . "Some of the scraped substance of the kidney
was gently agitated, in a test tube, containing a solution of albumen,
and a drop of this fluid was then placed under the microscope.
Thin sections were also used. In some animals no motion could be
perceived, but in the dog I observed currents taking place, in the
fluid, and also within the uriniferous tubes. The epithelial cells
would frequently disengage themselves from the sides of the tube,
and pass along for a considerable distance, and after emerging from
the mouth of the tube, would assume a rotatory motion. Sometimes
nearly all the epithelial cells would pass out of a tube, in the space
of fifteen or twenty minutes, leaving it almost denuded of its internal
epithelial lining. I have also seen isolated cells, having a vibratory
or rotatory movement. These appearances I have noticed upon

eight occasions, in the kidneys of dogs. Such motions generally cease, in these animals, in less than an hour after death. . . . I have often seen appearances similar to the preceding, in the kidneys of the ox and sheep. Nevertheless, it must be stated, that after very numerous and careful observations, I have never seen the epithelial cells actually provided with cilia, except upon one occasion, when I observed a single cell apparently fringed with cilia, and in active rotatory motion. This was in the kidney of the ox." . . . " What is observed in the kidneys of the dog, sheep, and ox, certainly seems to be much more powerful than the ciliary motion so readily seen in the oyster and clam, and very different in its nature from molecular motion. Reasoning from the appearances in the oyster and clam, as well as from analogy in many of the lower animals, it may be concluded that *ciliary motion does exist in those of a higher grade*, although it is, in them, very imperfect, or, as it might perhaps be said, in a rudimentary condition. In some of the inferior animals, where the urine is excreted in the semi-fluid state, a much greater necessity exists that this fluid should be rapidly propelled in its course through the uriniferous tubes, and, accordingly, we here find ciliary motion in its most perfect condition. I have never seen it in the human kidney, even after many careful examinations. If it does exist, which is probable, it ceases very soon after death, when it rarely happens that we can obtain an opportunity of examining it." Kölliker, in the last edition of his *Manual of Human Microscopical Anatomy*, Lond. 1860, states (p. 408) that "it (ciliary motion) is absent in birds and mammalia," and yet he has the title of Dr. Isaacs' paper, which was translated into Schmidt's Jahresber. in 1857, enumerated under the head of literature of the kidney. So that, as far as this distinguished histologist is concerned, Dr. Isaacs' observation of the fact of the existence of ciliary motion in the uriniferous tubes of the kidney in mammalia, is original.

In regard to the presence of epithelium upon the surface of the Malpighian tufts of the kidney, it is asserted by M. Morel, in the text, that he has seen it on several occasions in the kidneys of the guinea-pig, and also in man. This was originally asserted by Gerlach, but is doubted by Kölliker (*op. cit.* p. 408), and denied by Bowman and Dr. G. Johnson (*On Diseases of the Kidney*, Lond. 1852, p. 30, *note*). The observations of Isaacs upon this point are conclusive, as will appear from the following quotations from his paper. I will add also that I have myself witnessed several undoubted demonstrations of the fact at his hands.

"After much reflection as to the best plan of determining this point, I adopted the following processes : 1. By injecting watery and etherial solutions into the ureter, I succeeded in bursting the capsule, the Malphighian tuft, or coil, having been previously only slightly and, as was intended, imperfectly injected from the artery. Epithelial cells could then be seen upon the uninjected and transparent edges of the tuft, or coil. Plate 24, fig. 1, exhibits a Malpighian tuft. Broken fragments of the injected 'vessels are seen within it. They had been injected with chrome yellow, and appear black, when the specimen is viewed by transmitted light. The uriniferous tube had been distended by the injection from the ureter, and its expanded extremity or capsule had been burst, and can be perceived lying in shreds at the sides of the tuft, now uncovered by the capsule. Nucleated cells can be seen upon the naked and uninjected parts of the tuft. From the kidney of the black bear, magnified eighty diameters." In another specimen, the capsule had been scratched off with a needle, and then the naked tuft somewhat torn under the microscope. Some small fragments of the tuft could be seen with nucleated cells on their surface.

Again : "The capsule was torn off from a Malpighian body with a needle. In doing this, the capsule became reversed, so as to give a view of its internal surface, upon which small nucleated cells could be clearly and distinctly seen. *The surface of the naked tuft was covered by cells of much larger size than those upon the interior of the capsule. Upon the application of dilute nitric acid, the wall of the cells of the capsule was dissolved, while comparatively little effect was produced upon those of the tuft, thus showing a difference in their chemical constitution and organization."* This specimen was taken from the kidney of the racoon, and magnified 400 diameters.

Again : "Fine scrapings of the kidney of a cat were agitated occasionally for two or three days, in a test tube, the water having been frequently changed. By this method, the epithelial cells within the capsule were washed out, so that the space thus left between the tuft and the capsule became filled with water, which had soaked through the capsule. By slight agitation, while the specimen was floating in the water, under the microscope, it could be rolled over and over, so as to show various points of the surface of the tuft, *covered by nucleated cells.*"

" By the different processes first mentioned," he concludes, " I consider the existence of nucleated cells upon the surface of the Malpighian tuft, and, consequently, its analogy with the other secreting organs, as conclusively demonstrated." These descriptions are illustrated by several highly satisfactory drawings, pp. 404–407. The modesty of the author prevented him from emphasising the originality of the observation contained in the lines in italics, which were introduced by the writer. The importance of the anatomical fact, thus clearly demonstrated, that the Malpighian bodies of the kidney are covered by an epithelium, the cells of which are distinctly different from the ordinary epithelium of the urinary tubes, can be estimated by a reference to the fact that the ingenious and almost universally received theory of the action of the kidney, announced in 1842 by Mr. Bowman, in his paper in the Philosophical Transactions, is founded mainly on the belief that the surface of the vessels composing the Malpighian tuft were naked and bare. In a subsequent paper "*on the function of the Malpighian bodies of the kidney,*" read before the N. Y. Academy of Med., February 4th, 1857, its ingenious author, by a series of novel and interesting experiments, demonstrates, satisfactorily, that the function of the Malpighian tufts of the kidney is not the mere separation of water from the blood, as Bowman asserts, and that, on the contrary, *it separates from the blood most of the proximate elements of the urine,* any element of the urine which is not secreted by the Malpighian tuft, being, probably, afterwards separated by the epithelial lining of the tubes. *Trans. N. Y. Academy of Med., Vol. I., Part IX., p. 452.*

Another anatomical point settled by Isaacs is, that the ultimate ramifications of the renal artery do *not all* terminate in Malpighian tufts, although the great majority of them do so. At p. 385, he gives a wood-cut representing a small branch of the renal artery, as seen under the microscope, "which divides into two twigs, one of which supports at its extremity the Malpighian coil or tuft of capillaries, whilst the other enters into the venous plexus."

He also confirms, conclusively, the opinion of Bowman as to the relation existing between the Malpighian tufts and the expanded extremities of the convoluted tubes of the kidney, demonstrating, in opposition to the statements of Toynbee, Müller, Gerlach, Bidder, and others, "that the convoluted uriniferous tubes terminate by forming an expanded extremity, or capsule, which embraces the Malpighian tuft, or coil of capillaries."—(p. 401.)

" This difference of opinion, even among such high authorities, may probably be accounted for, inasmuch as it is very difficult, by employing the usual means of examination, to obtain any minute portion of the organ which will show the tube connected to the Malpighian body. Moreover, in examining any thin section of the kidney (as is usually done) for this purpose, it is to be remembered that the Malpighian body is always embraced in a ring of the *fibrous matrix*, and that at the neck of the tube, or its commencement of expansion into the capsule, is its weakest part, and where it is most easily, and indeed almost always, torn across, especially when traction is made upon it, as is usually done, in tearing out the specimen with needles. On the contrary, by using *scrapings* of the organ (as recommended) agitated in water, which softens and removes many of the adhering portions of the matrix, which last holds and confines the Malpighian body, this can be washed out, and not unfrequently with the convoluted tube attached to it." (See plates 15, 16, 19, 20, 21, and 22.) p. 404.

In relation to the '*fibrous matrix*' of the kidney, mentioned in the last quotation, nothing is said in the text. It is the more important to supply the omission, as this histological element of the organ plays a most important part in many of its diseased conditions. It was first described by Dr. John Goodsir, Professor of Anatomy, Univ. of Edinburgh, in the *Monthly Journal of Medical Science*, May, 1842 (see also Johnson *on Diseases of the Kidney*, Lond. 1852, pp. 16, 321). Its existence has been doubted by some high authorities, but the demonstrations contained in Dr. Isaacs' paper, and the numerous views which he gives of appearances presented under the microscope by his ingeniously prepared specimens, place all doubt at rest, and constitute him the highest authority on the subject.

The following quotations contain the essential points which he has demonstrated; but for a thorough knowledge of the subject, a perusal of his paper with its admirable illustrations is necessary. "I have never satisfactorily succeeded in exhibiting it (the matrix) by following the directions usually given for this purpose. It can, however, be always easily and distinctly shown by the following process: "Very thin slices of the kidney are to be made with Valentin's knife, put into a long test tube about one third full of water, and agitated from time to time for two or three hours. Prepared in this manner, a thin section exhibits under the microscope a kind of mesh, network, or honeycomb arrangement—the cells of the honeycomb, however, having no bottom. In the natural condition of the kidney, the smaller cells or openings transmitted the tubes; the large cells or openings were occupied by the Malpighian bodies; which last, together with the tubes, have now been washed out of the cells, although a few are often seen still remaining *in situ*, pl. 28." Here follow representations of similar preparations from the kidney of the rat, dog, rabbit, raccoon, hog, sheep, ox, horse, elk, moose, black bear, and finally of the human kidney. "According to my experience," he continues, "it is rare to find a human kidney which is *perfectly* healthy. This is particularly the case in subjects in the dissecting room, and in those who have died in large hospitals. Out of more than 500 subjects which I have examined in this city, I have seen but very few in which the kidney could be regarded as in an entirely healthy condition. Being very anxious to procure a perfectly healthy specimen of the organ, I obtained a considerable number of kidneys from the bodies of persons killed by violence and accidents, but these were also found to be diseased, most probably from intemperance, etc. I at length procured the healthy organs from which the present view of the matrix is here given."

"From what has now been said and exhibited, it is evident that the fibrous matrix is really the skeleton, or frame-work of the kidney. It consists of myriads of septa, or partitions, crossing each other in various directions, so as to form elongated spaces for the straight tubes, or rounded spaces, cells, or rings, for the Malpighian bodies and convoluted tubes." "In certain pathological conditions of the kidney, it becomes diminished in size, and indurated; its surface is irregular and covered with small projecting points, like variously sized shot. This condition is not unfrequently seen in old drunkards, and would seem to be analogous to cirrhosis of the liver, and probably induced in a similar manner; the alcoholic fluid passing through the tubes of the kidney, and by continued irritation producing crisping, or irregular contraction of the meshes of the matrix, which consequently constrict the tubes and the Malpighian bodies."

" Thickening and induration of the matrix may produce injurious effects otherwise than by constriction of the tubes and Malpighian bodies. The minute vessels and capillaries pass through the substance of the fibrous rings. Consequently, induration and contraction of the matrix, by direct pressure on the vessels, must greatly interfere with the circulation, nutrition, and secretion of the kidney, and thus various morbid products, as blood, albumen, pus, tubular casts, etc., may be found in the urine." Pp. 418-429. By isolation of portions of the matrix by repeated washings, and the subsequent application of dilute acetic acid and other reagents, it was found to consist entirely of fibres, containing elongated fusiform nuclei; in other words, to be pure connective, or white fibrous tissue—the yellow elastic element being invariably absent.

" From the foregoing remarks, it is evident that a correct knowledge of the fibrous matrix is of great importance, and that its microscopical and chemical investigation, in all cases of diseased kidney, would probably furnish interesting and valuable results." P. 431.

It remains to notice the original and very ingenious methods of preparing the specimens by means of which Dr. Isaacs succeeded in attaining such successful results. His aim was to render the substance of the kidney transparent under the microscope, and to effect this he instituted numerous experiments with chemical reagents, and " at length arrived at the knowledge of certain processes, which have not only been useful by giving transparency to small portions and thin sections of the organ, but have often enabled them to be viewed both as opaque and as transparent objects." With these processes, all the usual means, such as injections, etc., were also employed. The following, in addition to those already indicated, are the modes of preparing microscopical objects which he praises most highly.

To show the epithelium of the urinary tubes, their sections, or *scrapings* of the kidney, should be kept in a solution of albumen in fresh urine. " To view the tubes of the kidney in their normal condition very thin sections, or scrapings of the cut surface of the organ, may be put into a test-tube with water, agitated for a few minutes, placed on a slide of glass, covered by a thin slip, and then examined under the microscope." Or, to render the object transparent, scrapings are put into a test-tube with about half an ounce of water, to which three drops of pure sulphuric acid are added, and the whole boiled for one or two minutes. " If too much acid be used, it will dissolve all the minute vessels and capillaries, but not the Malpighian coil, or tuft. This is worthy of notice, as showing a great difference in the chemical constitution, and consequently in the organization, of these different parts."

The addition of chloroform, under similar circumstances, also has the effect of rendering objects transparent. " I have succeeded in exhibiting the tubes of the kidney very distinctly, by boiling small pieces of the organ in diluted chloroform, and also in solution of chlorate of potassa, which latter is useful in giving a clear view of the surface of the kidney when congested, as it shows the venous plexus and the tubes, and acts but slowly upon the blood-globules." To show the vessels in connexion with the Malpighian bodies, and at the same time to exhibit the tubes :—after injecting the vessels with white lead finely ground in oil (artists' tubes), and well agitated with sulphuric ether, small pieces of the organ were boiled in very diluted chloroform, and some thin sections were made with Valentin's knife ; other sections were first dried, then immersed in spirits of turpentine, and finally placed on a glass slide, in a drop of water, under the microscope. " Muriatic, acetic, and nitric acids, also exhibit the tubuli with considerable distinctness; the last, however, is apt to discolor the tubes. I have also tried the effect of many other chemical reagents; among others, the phosphoric, chromic, boracic, tartaric, and citric acids, the alkalies and their carbonates, various salts, etc."

In conclusion, I must again urge the student who wishes to fully comprehend the anatomy of the kidney, to consult this admirable paper, and study its graphic illustrations ; it is the ablest anatomical monograph that our country has as yet produced, and a fit contribu- ·tion, by its lamented author, to the science to which he devoted his life.]—(*Ed.*)

Testicle. *Testicle.*—The proper coat of the testicle (*tunica albuginea*) is of the same structure as that of the kidney—but somewhat more dense. Its external surface, excluding that portion corresponding to the hilus of the organ, is clothed with a simple layer of the epithelium, which constitutes the visceral lamina of the serous membrane of the testis (*tunica vagina-* structure. *lis*); its internal surface is in immediate relation with the secreting tubules of the gland. Along the line where its vessels enter and leave the organ (hilus) the *tunica albuginea* presents a very considerable thickening (*corpus Highmorianum*), which, in the form of a ridge, buries itself in the parenchyma of the gland, and from its surface the interlobular fibrous septa take their origin. In the interior of the *corpus Highmorianum* is a tubular network (*rete testis*), from which the secreting tubules are given off on one side, and from the other, the efferent canals which terminate in the excretory duct.

From the interior of the *rete testis* twenty to thirty vasa recta. straight tubes, $\frac{1}{75}$th of a line in their mean diameter, take their origin (*vasa recta*), and, after a short course, divide into several branches. Each one of them becomes exceedingly tortuous (Pl. XXI. fig. I), and again gives off several subdivisions, which anastomose with each other, and terminate either by loops, or by blind extremities. Sometimes one of these branches, but more frequently two or three of them closely united, form a lobule of a conical shape, the apex of which is in relation with the *vasa recta*, whilst its base looks towards the periphery of the organ. Not unfrequently canals of communication exist between these lobules.

The upper extremity of the *rete testis* gives off a Efferent tubes.

dozen tubuli (*vasa efferentia*) which, by their union, constitute the head of the epidydimis (*globus major*). Immediately after emerging from the *corpus Highmorianum* these tubes become exceedingly convoluted, and proceed in succession to the *globus major* of the epidydimis; each of them takes the shape of a cone (*coni vasculosi*), the apex of which is continuous with the *rete testis*. Finally, the canal of the epidydimis, after having by its convolutions formed the body of the epidydimis and its lower extremity (*globus minor*), and after giving off the *vasculum aberrans*, becomes the *vas deferens*, or excretory duct of the gland.

Tubuli testis. The spermatic tubules (*tubuli testis*) and the *vasa recta*, which are directly continuous with them, have very thick walls, consisting of several distinct layers. The external, which is the thickest ($\frac{1}{757}$th of a line), is fibrous and very rich in plasmatic cells (Pl. XXI. fig. II. 1); the internal, exceedingly thin and structureless (fig. II. 2), is in relation, by its inner surface, with the stratified epithelium by which the cavity of the tubule is completely filled (fig. II. 3). The cells of this epithelium are of considerable size, polyhedral, and, in the adult, most of them contain oil-globules. In the *rete testis* the tubuli have no proper walls; they are simply tunnels through the fibrous substance of which the corpus Highmorianum is composed.

Epidydimis and vas deferens. In the epidydimis the non-striated muscular element is added to those already recognised in the walls of the tube, and its epithelium is cylindrical (Pl. XXI. fig. III.). The *vas deferens*, proceeding from without inwardly, is composed: 1st. of a fibrous coat; 2d. of a muscular coat, with both longitudinal and circular fibres ; 3d. of a mucous coat, lined by simple pavement epithelium. The *vesiculæ seminales*, with the exception that their walls are thinner, possess the same structure as the vas deferens.

The arteries of the testes, ramifying in the inter- Vessels and nerves. lobular septa, terminate in a capillary plexus which surrounds the *tubuli seminiferi.* The veins follow the course of the arteries. The lymphatics are very numerous ; they accompany the vessels of the cord, and run into the lumbar glands. The nerves, few in number, follow its arteries into the parenchyma of the gland, but how they terminate there is not known.

The *semen* is an alkaline liquid, destitute of color, Seminal fluid. in which certain accessory elements are encountered, such as cells and the debris of cells, and also certain essential and characteristic elements—the filiform corpuscles endowed with the power of motion, and known under the name of *spermatozoa.* These curious corpuscles consist of exceedingly delicate filaments with an almond-shaped enlargement at one extremity, constituting its head, whilst the remainder of the filament, tapering off to an extremely fine point, represents the tail. A delicate linear depression, or species of collar, marks the line of junction of these two portions, on the borders of which a minute tubercular projection is sometimes to be seen (Pl. XXI. fig. IV.). No trace of organization can be detected, on the closest examination, in any part of the spermatozoa ; the substance of which it is composed being homogeneous, transparent, and amorphous. The movements of the spermatozoa continue a long time after death (10 to 24 hours); water and acids arrest them, but they are renewed again by the application of slightly alkaline liquids.

Development of spermatozoa. These elements, which constitute the essential portion of the seminal secretion, are developed in the following manner: On examining the epithelium of the *tubuli seminiferi* it is observed that its central cells are generally more voluminous than the rest, and give evidence of a more or less active process of endogenous vegetation (Pl. XXI. fig. V.); thus, some cells are to be seen enclosing as many as ten nuclei. In addition to the nucleolus, which is visible in each nucleus, in the shape of a brilliant spherical vesicle, there is also observable upon one point of its periphery an elongated spot (fig. V. 3), from which shortly a long and delicate filament takes its origin, which coils upon itself as it increases in size, occupying always the periphery of the nucleus (fig. V. 7). As soon as the nuclei have attained their full development, the parent cell, being no longer able to contain them, bursts, and they thus become free. Its elements subsequently become disintegrated and disappear. The tail of the spermatozoon then uncoils itself (fig. V. 8), afterwards its head becomes disentangled, and its development is accomplished.

In the testes of the guinea-pig, we have, on several occasions, traced the successive transformations which the epithelial cell undergoes in giving origin to spermatozoa, and have always witnessed the occurrence of the phenomena taking place as above described; that is, each nucleus distinctly gives origin to a spermatozoon.

In the human testis the same process has been demonstrated.

The testicle is developed from the internal substance of the Wolffian body, whilst its excretory duct is formed by the external canal of the same organ The *vas aberrans* takes its origin from the centre of the Wolffian body, from which, according to some authorities, the epidydimis also is formed. The histological changes which take place during the development of the testicle are not demonstrable with sufficient certainty to justify their formal description.

Development the testicle.

What has been said already in reference to the epithelium, and mode of secretion, of glands composed of clusters of follicles, applies with equal force to those composed of tubes. Those which possess a single layer of epithelium, and in which secretion is effected by simple filtration, are: the glands of the intestinal canal, those of the uterus, the sweat glands, the kidneys, and the liver.

The glands which have a stratified epithelium, and in which secretion is effected by vegetation of its cells, are the ceruminous glands, and the testicles.

SECT. III. MIXED GLANDS.—The ovary in some respects resembles the blood-glands, but differs from them by possessing an excretory canal, and by the peculiar character of its follicles; the liver, with its double apparatus for the secretion of bile and sugar, is both a tubular and a blood-gland. The structure of these two organs makes it necessary, therefore, to associate them together in a separate group, which constitutes naturally a connecting link between the true and blood-glands.

Ovary structure.   *The Ovary.* The envelope of the ovary (*tunica albuginea*) is of the same nature as that of the testicle, and like it, is covered everywhere, except at its lower border (*hilus*), by a serous membrane. Its inner surface is intimately adherent to, and continuous with the parenchyma of the organ. This is composed of an obscurely fibrous substance, traversed by numerous blood-vessels, in the meshes of which the *ovisacs*, or *Graafian vesicles* are found. At the lower border of the ovary its fibrous element is denser than elsewhere, and forms, with the *tunica albuginea*, a sort of *corpus Highmorianum*, in which there are no ovisacs ; but just outside of it, and throughout the whole parenchyma to its outer surface, large numbers of them exist, and of all dimensions—the larger ones always lying nearest to the surface of the organ.

Ovisac.   The outer envelope of the ovisac is a fibro-vascular membrane consisting of the same material as the parenchyma in which it is embedded, only more condensed. The external portion of this membrane, which is loosely adherent to the surrounding tissue, is less vascular than its deeper surface. In contact with this is a stratified epithelium (*membrana granulosa*), which increases considerably in thickness near the point which looks towards the surface of the ovary, and constitutes the *proligerous disc*, in the interior of which the ovule or egg is contained (Pl. XXI. fig. VI. 2, 3, 4). The remaining cavity of the ovisac is filled by an albuminous fluid which contains cells, or the debris of cells, detached from the *membrana granulosa*.

Ovule.   The proper tunic of the ovule (vitelline membrane,

zona pellucida) is about ₁₅₀th of a line in thickness,
transparent, and entirely structureless. Its contents,
the vitellus or yelk, scarcely liquid in consistence,
contains a great number of very fine granules, probably fatty in their nature. At one point of the
periphery of the vitellus there is a brilliant spherical
nucleus, the *germinal vesicle*, or vesicle of Purkinje.*
Finally, the nucleus itself contains a nucleolus, called
the *germinal spot* (Wagner).†

Generally, as is well known, the ovisac contains but
one ovule; nevertheless, it may contain two, as we
have seen in one instance, in the ovary of an adult (Pl.
XXI. fig. VII. 1, 2). We have also observed an
example of segmentation of the vitellus in another
ovum of the same female, which proves that this phenomenon may take place in the interior of the ovary
without previous fecundation (fig. VIII).

The Fallopian tube, or oviduct, which serves as the   Fallopian tube.
excretory duct of the ovary after the detachment of
the ovum, has three layers of tissue in its walls; the
first is serous, and belongs to the peritonæum; the
next is composed of smooth muscular fibres and
blood-vessels; the third is mucous membrane, the
surface of which presents longitudinal plaits, and is
covered by a single layer of cylindrical ciliated epithelium. The vibratory motion of the cilia tends to
carry the contents of the tube onwards from without
inwards, and consequently facilitates the descent of
the ovum towards the cavity of the uterus.

---

* Purkinje, Professor of Physiology in the University of Prague.—(*Ed.*)

† Rudolf Wagner lately resigned the Chair of Physiology at the University of Gottingen, and was succeeded by Meissner.—(*Ed.*)

Uterus.   The uterus, into which the ovum passes from the
Fallopian tube, possesses the same tissues in its walls,
only the second, or muscular coat, is much more
developed than in the oviduct, especially during gestation.   Its third, or mucous lining, is also more complicated in its structure; the portion of it which
belongs to the body of the organ resembles that of
the Fallopian tube, but that which lines its neck is
studded with filiform papillæ which overlap each
other; they are found near its inferior orifice; in
addition, we have, in the mucous lining of the uterus,
large numbers of tubular glands, which have already
been described.

The vagina has also three coats: the outer, fibrous;
the middle, musculo-vascular; and internal, mucous.
On the latter we find numerous deep rugæ, especially
towards its orifice, and a large number of conical
papillæ.   Its epithelium is thick and stratified.   Thus
far no glands have been found in the thickness of its
walls.

Vessels and   The arteries of the ovary enter that organ at its
nerves.
lower border, and break up into a great many very
tortuous branches, of which some are distributed to
its parenchyma and fibrous envelope, whilst the rest
go to form a close capillary plexus in the walls of the
ovisacs.   Its veins follow the track of the arteries,
and terminate in the ovarian and uterine veins.   Its
lymphatics, whose origin is not fairly made out,
accompany the blood-vessels, and finally communicate
with the pelvic and lumbar glands.   As to its nerves,
they run with its arteries into the secreting portion
of the organ, but their distribution and mode of termination are unknown.

The ovary is developed at the expense of the inte- rior substance of the Wolffian body, and is distinguishable ·from a testicle, by not being continuous with its excretory duct. It consists, at first, exclusively of embryonic cells, the greater part of which are transformed into the ovisacs; the remainder form the parenchyma and envelope of the organ. In the ovaries of the fœtus, or newly born infant, the ovisacs are readily recognised as a series of pockets lined internally by a layer of epithelial cells which surround another larger central cell; this eventually becomes the ovula, whilst those which surround it, increasing in number, form the *membrana granulosa*, and proligerous disc.

The body of Rosenmuller, or *parovarium*, situated between the layers of peritonæum which connect the ovary and fimbriated extremity of the Fallopian tube, and consisting of a number of blind canals, represents the remains of the central portion of the Wolffian body, and corresponds to the *vasculum aberrans* in the male.

The *liver* is covered externally, on all sides, by a layer of connecting tissue which, at its transverse fissure, applies itself to the vessels of the organ, and following them into its parenchyma, accompanies them in their ramifications, forming a common investment for them, and the whole organ, which is known as the *capsule of Glisson*.

The outer surface of this proper coat of the liver is very closely united to the peritonæum throughout its whole extent, except at its posterior border, its transverse fissure, and at the fissure for the gall-bladder.

Its internal surface sends innumerable delicate processes, or trabeculæ, into the interior of the organ, which traverse its parenchyma, and become ultimately continuous with the connecting tissue which accompanies its vessels to their extreme ramifications.

The parenchyma of the gland consists of an aggregation of little granules, one-fourth to one-half of a line in diameter, of a yellowish-brown color, and darker centrally than on the surface. This color varies considerably, for it depends upon the amount of blood in the ramifications of the portal vein, and in those of the hepatic vein, and also upon the degree of fatty infiltration which exists in the hepatic cells. In the human liver the outlines of these granules, called also *acini* or lobules, are rather indistinct; but in the hog it is different; here each lobule forms a little polygon, entirely independent of its neighbors, and presenting a very distinct outline.

Lobules. Each lobule of the liver represents, in its structure, a miniature of the whole organ; it will suffice therefore to study one of them thoroughly, in order to get an exact idea of the histology of the gland. In each lobule we find: 1st, a mass of hepatic cells; 2d, the terminal ramifications of the venæ portæ and of the hepatic veins; 3d, the biliary apparatus, consisting of the radicles of the hepatic duct, and the terminal branches of the hepatic artery.

The cells are of two sorts: the one, which constitutes pretty much the whole epithelial mass, are large and irregular polygons, $\frac{1}{70}$th of a line in diameter, with nuclei almost always infiltrated with fat, and contents consisting of free oil-globules, and a large quantity of minute pale granules, which Schiff[*] regards as a species of animal starch (Pl. XXII. fig. IV. 1). The other cells, much smaller ($\frac{1}{3000}$th of a line) and less numerous, have a regular polygonal form, and finely granular contents, generally free from fat (fig. IV. 2). In the interlobular spaces, which in the hog's liver are very distinct, we find minute branches of the venæ portæ (interlobular veins), by which the lobules between which they are situated are supplied with blood (Pl. XXII. fig. I. 2, 3). If we examine closely the relations between the interlobular veins and a single lobule, we find them forming around it a vascular network, from the concavity of which a multitude of minute ramusculi take their origin, which penetrate the substance of the lobule, and immediately break up into capillaries (fig. I. 4; fig. II. 2). They anastomose freely with each other, and thus form a network with very close meshes ($\frac{1}{70}$th of a line), which empties, finally, at the centre of the lobule, into a minute veinule, which is a radicle of the sub-lobular hepatic vein (fig. III. 2). In man, the lobules not being clearly limited in their outlines, the vascular circle thus formed by the penultimate branches of the portal vein is not readily distinguished, but the capillary network in their interior has the same appearance and arrangement as that described (fig. II.).

The large hepatic cells, usually connected to each other in pairs, fill up the meshes of the capillary plexus of the lobule, and in their aggregate mass con-

* Professor of Physiology in the University of Zurich.—(*Ed.*)

stitute an epithelial network, which is interwoven thus with the web of capillary vessels. The elements of the liver which we have studied thus far constitute its *glycogenous* apparatus.

Hepatic duct.   The hepatic duct enters the liver at its transverse fissure along with the hepatic artery and portal vein, and accompanies these vessels to the lobules. In its course it gives off a great many arborescent branches, of which the larger anastomose frequently with each other, whilst the smaller ones remain solitary and continue on to the surfaces of the lobules. From these minute perilobular ramifications arises a set of capillary tubes which enter the substance of the lobules—not very deeply—and there terminate in blind extremities; they are lined within by a simple epithelial layer, made up of the smaller cells which have been described above (Pl. XXII. fig. VI. 5, 6, 7). In the biliary ducts which possess a diameter beyond $\frac{1}{50}$th of a line, according to Kölliker, their pavement epithelium is replaced by a cylindrical epithelium. Finally, the hepatic ducts proper, together with the cystic duct and the *ductus communis chole-dochus* have muscular fibres in their walls, and also a considerable number of minute racemose glands. The hepatic artery accompanies the hepatic ducts to the lobules, supplies them with very numerous branches, and finally terminates in the capillary plexus of the portal vein. This second apparatus, composed of the biliary ducts and the hepatic artery, constitutes a tubular gland. The liver then is made up of two glands, which are intimately intermingled with each other; one of these (a blood-gland) is concerned in the secretion of sugar, and the other (a tubular gland) in the secretion of bile. The physiology of the liver, so well established by M. Claude Bernard, as well as its comparative anatomy, entirely justify this view.

The mode of origin of the radicles of the hepatic duct has been the object of varied researches, and the number of theories which have been proposed in explanation of it are evidence of the uncertainty which surrounds the subject. We have admitted that the liver is made up of two distinct glands, because, as we have already said, both physiology and comparative anatomy unite to prove it, and we have described the radicles of the hepatic ducts as tubes terminating in blind extremities, because we have witnessed this fact twice in a cirrhotic liver, and we know that Professor Küss* had already observed the same several years since in a syphilitic liver.†

* Professor of Pathological Anatomy in the Faculty of Strasbourg, France.—(*Ed.*)

† It will be observed that our author, in his very succinct account of the minute anatomy of the liver, makes no allusion to the labors of Kiernan, Lereboullet, or Beale, an omission which cannot be fairly overlooked. These authors throw too much light upon the much-debated question of the mode of origin of the radicles of the hepatic duct to be passed over in silence. The best supported opinion on this subject at the present day is not that the radicles of the hepatic ducts are merely tubes terminating in blind extremities, as stated in the text, but that these radicles take their origin from a *biliary plexus* or network of tubes, analogous to the network of blood-vessels in the interior of the hepatic lobule, and occupying its interstices. Kiernan first described and figured this "biliary plexus" in his admirable paper on "*the Anatomy and Physiology of the Liver,*" in the Philosophical Transactions, London, 1843, (part 1st, p. 741, and Pl. XXIII. fig. III). His figure, however, is diagrammatic, and not taken from nature. It has been reproduced in most of the works on Descriptive Anatomy of the present day.

In 1853, M. Lereboullet published a prize essay, in Paris, "*on the Mi-*

The walls of the gall-bladder are composed of the same elements as those of the larger biliary ducts. Its mucous membrane has numerous folds and rugæ

nute Structure of the Liver," in which, after very thorough investigation of the subject, he arrives at the conclusion that Kiernan's hypothesis is correct, and that the biliary plexus is truly demonstrable, although, after long and laborious research, he was not able to demonstrate the existence of the membrane forming the walls of the tubes composing it. He speaks, however, with confidence, of the network of tubes, or biliary plexus, containing the secreting cells of the liver (hepatic cells), as occupying the interstices of the vascular plexus of the lobule—so that the two sets of tubes, biliary and vascular, interlace with each other accurately, the walls of the bile-tubes being everywhere so closely adherent to the external surfaces of the blood capillaries that the former were not susceptible of separate demonstration (p. 59 et seq.).

In a Monograph by Dr. Lionel S. Beale (*on some points in the Anatomy of the Liver of Man and Vertebrate Animals*, London, 1856), the following passage occurs in the summary of his able and interesting investigations on this subject:

" The liver cells lie within a tubular network of basement membrane, which separates them from the walls of the capillaries. In many cases, however, these thin membranous tubes cannot be separated, and are, no doubt, incorporated with each other."

" The cell-containing network is directly continuous with the most minute ducts, which ramify at the circumference of the lobule, and it terminates in the centre by loops, which lie close to the intralobular vein." (p. 74.)

Both Lereboullet and Beale describe the tubes composing the " cell-containing network" as larger in diameter than the "minute ducts," which are the radicles of the hepatic duct. " The tubes of the cell-containing network are about $\frac{1}{1000}$th of an inch in diameter, or more, but the finest ducts are commonly not more that $\frac{1}{3000}$th, and they are often seen even less." . . . " The smallest ducts are lined with a very delicate layer of epithelium, composed of *flattened cells* of a circular form, contrasting remarkably with the large *secreting cells*"—the hepatic cells proper, which occupy the cell-containing network, or biliary plexus. (Beale, p. 74.)

In addition to the evidence we have quoted, which sets forth what we believe to be the true anatomy of the liver in relation to the point in question, the monographs of Lereboullet and Beale may be advanta-

upon its surface, the intersection of which gives it a
honey-combed appearance; it is lined by a cylindrical
epithelium, composed of yellowish colored cells, which
are very pale and often destitute of nuclei (Pl. XXII.
fig. V.).

The liver has two sets of lymphatics: one super- Lymphatics.
ficial, ramifying in the thickness of its proper coat;
the other deep, and following the subdivisions of the
portal vein; they anastomose freely with each other,
and terminate, those from the convex surface of the
organ, by passing upwards through the thoracic
cavity, and those from its inferior surface, by running
into the lymphatic glands of the abdomen.

The nerves of the liver, which are derived from the Nerves.
pneumo-gastric and great sympathetic, join the hepatic artery and follow its ramifications; but their mode
of termination in the interior of the lobules is unknown.

The first traces of the liver make their appearance Development.
in the shape of two little masses of cells, one upon
the outer, the other upon the inner or epithelial layer
of the intestinal wall. As to their subsequent metamorphoses, what has been ascertained by the most
reliable embryologists may be stated as follows:
Whilst the more external of the two cellular masses
enlarges and surrounds the common trunk formed by
the umbilical and portal veins, constituting thus a
parenchymatous mass enclosing the portal system of
veins which are closely enveloped by large sized

geously consulted by the student; they contain much valuable information as to injecting and preparing specimens of the liver for microscopical study.—(*Ed.*)

hepatic cells, the internal mass sprouts into numerous tubular branches which, penetrating the substance of the outer mass, constitute the system of biliary ducts. Whilst these changes are taking place, the original primordial cells, of which the masses were composed, undergo various metamorphoses which result in the formation of the several tissues constituting the substance of the gland.

Preparations.

The preparations on which the structure of the liver is best studied are very delicate sections of the fresh liver, and also of livers injected with colors ground in oil and diluted with essence of turpentine.

Ductless glands of the intestines.

SECT. IV. DUCTLESS FOLLICLES AND BLOOD GLANDS. —The most simple in their structure of this class of organs are the solitary glands of the intestinal canal. These are little spherical bodies, situated deeply in its mucous membrane, and varying from half a line to one line in diameter. Their walls consist of fibroid membrane studded thickly with plasmatic cells. Their contents, greyish in color, and of a solid consistence, are made up of a quantity of rounded globules, from $\frac{1}{300}$th to $\frac{1}{150}$th of a line in diameter, and bearing a very close resemblance to the cells contained in the interior of the lymphatic glands. Numerous blood-vessels ramify upon their exterior, and then, penetrating their walls, converge towards the centre of each follicle. Externally to the follicle they anastomose freely, but towards its centre they form a great number of loops, which, when filled by injection, present a very beautiful appearance (Pl. XXVI. fig. XIII. 2). Kölliker has found nerves in the follicles of the tongue, and, according to this author, E. Weber has also recognised in them radicles of lymphatics. No external outlet has been discovered, after the most rigorous examination of the external surfaces of these glands. Reasoning from the structure of these organs, and also from certain pathological relations existing between them and the mesenteric glands, the conclusion seems to us legitimate that they are identical in their nature with lymphatic glands. This opinion is sustained, also, by the most distinguished histologists of Germany. These ductless glands are what are known as the solitary glands of the intestine ; they constitute the essential portion of Peyer's glands, of the tonsils, the follicles of the tongue, and the pharynx. The structure of the vesicles of the thymus gland warrants us also in including it in the same category.

Thyroid body.—The fibrous envelop of the thyroid Thyroid body body gives off from its internal surface a great number of delicate trabeculæ of connecting tissue by which its interior is divided up into spaces, or cavities, in which the vesicles of the gland are contained (Pl. XXII. fig. VIII.). Each vesicle is thus enclosed by a delicate partition of connective fibres, in the substance of which a considerable number of plasmatic cells are found, together with an abundance of vessels (fig. VIII. 2, 3). A single layer of polygonal epithelium lines the interior of the vesicle, and its cavity is filled with an albuminous fluid. In the fœtus, and young child, the epithelial layer consists of cells $\frac{1}{1000}$th of a line in diameter, with finely granular contents, and a nucleus measuring about $\frac{1}{1600}$th of a line. But in the adult, and in old age, it is very

9

difficult to find the epithelium well marked and distinctly separate from the liquid contents of the vesicle; most frequently its cells are found infiltrated with fat, and a large number of free oil globules and nuclei are contained in the liquid which fills the vesicle, which is evidence of the disintegration and breaking down of its elements (fig. VIII. 4). It is doubtless in these vesicles, or follicles, of the thyroid, that most of the morbid growths to which the gland is liable, take their origin. The lobulated shape of the thyroid body depends upon the aggregation of its follicles into little masses, or lobules, which remain to a certain extent independent of each other.

Vessels. The blood-vessels of the thyroid are remarkable for their size and number, and the rich and delicate plexuses of capillaries which they form around the walls of each follicle. Its nerves come from the sympathetic; with regard to their mode of termination, as well as to the distribution of its lymphatics, nothing is certainly known.

Development. The facts ascertained thus far·in relation to the development of the thyroid gland are not sufficiently accurate and positive to justify their record in this work.

Spleen. *Spleen.*—The spleen is an organ in regard to the structure of which much remains yet to be elucidated; the following account includes all that the labors of the most eminent microscopic anatomists have thus far established as certain.

The envelop of this vascular gland is similar to that of the liver, and is intimately adherent, externally, to the peritonæum, and internally, to its parenchyma ; at the *hilus* of the organ it is reflected upon its vessels, and accompanying them into its interior, forms a sort of capsule of Glisson.

Its parenchyma consists of a sort of *corpus caver-* Parenchyma. *nosum*, the trabeculæ of which, fibrous in their nature, are continuous by their extremities with the inner surface of the proper coat of the organ, and in their interspaces is a soft material very closely resembling blood-clot, called the pulp of the spleen. There are also certain spherical bodies connected with its arteries, and called corpuscles of Malpighi, which form a portion of the parenchyma of the organ.

We have said that the fibrous coat of the spleen was reflected upon its vessels, and constituted for them a sheath or capsule, continuous, in its interior, with the trabeculæ. The splenic artery gives off, near its hilus, a certain number of branches, which, as they penetrate the substance of the gland, continue independent of each other, and form, each one of them, by its subdivisions, a sort of vascular brush. Situated upon arterioles of from $\frac{1}{20}$th to $\frac{1}{15}$th of a line Malpighian in diameter are the little rounded whitish colored bodies. bodies already mentioned as the corpuscles of Malpighi ; they measure from one-eighth to one-fourth of a line in diameter. Their structure is identical with that of the solitary glands of the intestinal canal. They have a very delicate outer coat of connecting tissue, which is continuous with the sheath of the artery upon which each corpuscle is situated, and its contents are made up of the same elements that we find in the interior of a lymphatic gland, viz., spherical cells of from $\frac{1}{800}$th to $\frac{1}{600}$th of a line in diameter, and free nuclei ; Kölliker mentions also the presence of blood globules, both normal in appearance, and in various stages of alteration. Leydig has described a vascular network which penetrates the interior of the corpuscle, and thus completes its resemblance to a ductless gland ; this author, in fact, does not hesitate to consider the Malpighian corpuscle of the spleen as a minute lymphatic gland.

Splenic pulp. The splenic pulp, the substance of which is traversed by the finest vessels, and the most delicate trabeculæ, includes elements of different kinds. Its principal bulk consists of cells similar to those of the Malpighian corpuscles (Pl. XXII. fig. VII. 1) ; debris of red blood globules, blood pigment, and larger cells with many nuclei, $\frac{1}{70}$th of a line in diameter, make up the remainder. In some of these latter elements the formation of red globules seems to take place by endogenous vegetation; at least this is to be inferred from the drawing by Otto Funke,* in his Atlas of Physiological Chemistry. But Kölliker asserts, on the contrary, that these cells are at first nothing more than aggregations of red globules, around which a membrane has formed, thus constituting a cell, of which these old blood globules become the nuclei, and that they are about to undergo still farther metamorphoses in a retrograde direction, of which he gives drawings confirmatory of his opinion. Are we to conclude then, with Otto Funke, that red globules of the blood are formed in the

* Formerly Professor of Physiology in the University of Leipzig; at present occupies the same chair at Freiburg, Grand Duchy of Baden, Germany.—(*Ed.*)

spleen, or, with Kölliker, that it is an organ whose function is to destroy them ?

Amongst the elements of the spleen there remain to be noticed some cells of a fusiform shape, very much enlarged around their nuclei (fig. VII. 2). In comparing these corpuscles with the epithelial cells of the vessels, it is difficult, from their similarity of shape, not to regard them as identical in nature Nevertheless, Führer* (Gazette hebd., 1855, p. 314), takes a different view of them, and assigns to them an important physiological function. According to his view, these fusiform cells, swelled out like so many aneurisms, are nothing more than tubes connected to each other, end to end, and communicating with the capillaries of the spleen, whilst their nuclei are the future red blood globules. Führer has doubtless allowed himself to be carried away by the desire to establish an analogy between these fusiform bodies and the arterial dilatations which exist in fishes, and he has overlooked their epithelial character.

The smallest of the veins spring from the capil- Veins, etc. laries, run alone for a short distance, and then join the arteries. The lymphatics, superficial and deep, meet and unite at the hilus, whence they are transmitted directly to the thoracic duct. The nerves of the spleen are numerous; they are distributed with its arterial branches, and seem to terminate by free extremities.

It is a matter of extreme difficulty to determine the relations which exist between the splenic pulp,

* A practising physician at Hamburgh, author of a popular work on Surgical Anatomy.—(Ed.)

and the larger veins of the organ. Is the pulp entirely outside of the vessels, as asserted by some authorities, or are the spaces in which it is contained simply dilatations of the veins, and if this be true, does it form a part of the general circulation? The researches which seem to us most conclusive tend rather to establish the entire independence of the cavities which contain the splenic pulp; yet the fact is not to be accepted as demonstrated.

Development.   It is not agreed at what point exactly the spleen takes its origin. Arnold asserts that at first (from the seventh to the eighth week), it is confounded with the pancreas. Bischoff* says that, in the fœtal calf, he has seen it spring from the greater curvature of the stomach; and according to other observers, it is developed from a blastema, at first isolated, but which shortly becomes attached to the great cul-de-sac of the stomach. As to the histological transformations which the organ undergoes during its increase of size, they have not been accurately demonstrated.

Supra-renal capsules.   *Supra-renal Capsules.*—The nature and physiological function of these organs are as yet unknown. Nevertheless it seems probable, from the nature of certain elements which enter into their composition, and which form, in fact, almost the whole of their central portion, that they belong rather to the nervous system than to the class of glands under consideration.

Structure.   The supra-renal capsule possesses a thin envelop, fibrous in its structure, and intimately connected with

* Professor of Physiology at Heidelberg, Baden, Germany.—(*Ed.*)

the parenchyma of the organ by delicate processes, or trabeculæ, given off from its internal surface.

On making a section completely through the body of the organ it is found to consist of two distinct substances, of which one forms its cortical portion, and the other its centre. The first, or cortical substance, one-half a line to a line in thickness, is of a soft, solid consistence and brownish color, being of a somewhat deeper tint externally than internally. In the fibrillated tissue which forms its basis are a number of elongated cavities filled with large cells (from $\frac{1}{100}$th to $\frac{1}{60}$th of a line in diameter) and very much infiltrated with fat.

The fibrillated element of its central or medullary substance is more delicate than that of its outer portion, and the cells which it contains resemble exactly the multipolar or caudate cells of the nervous ganglia. The numerous nerves with which the suprarenal capsule is supplied penetrate this substance, and unite, as Leydig has demonstrated, with the prolongations of these nerve cells. We have a right, then, in accordance with these facts, to regard this organ as a nervous centre, and to consider its cortical layer as simply a protecting membrane, for the very considerable amount of fatty infiltration of its cellular element seems to indicate an arrest of its functional activity.*

* In one of the most aggravated instances of hysterical temperament that I have ever encountered, the patient, a maiden lady, died, at the age of 84, of cancer. The cancerous deposit involved both supra-renal capsules, which were each as large as the closed fist; the right, which was somewhat the larger, having imbedded itself firmly in the under surface of the liver. There was also a cancerous mass in the posterior wall of

Vessels.   Its arteries and veins are numerous; in their mode of distribution they present no peculiarities worthy of note. ʸThe lymphatics are few in number, and seem to belong to the cortical substance alone. The development of the supra-renal capsule takes place at the same time with that of the kidney, but entirely independently of that organ. Of the histological transformations which it undergoes during fœtal life little is known.

the transverse colon, and another in the root of the right lung. This latter, by its pressure upon the pulmonary veins, the interior of which it had also invaded, gave rise first to hæmoptysis, and afterwards to extensive pleuritic effusion—from the immediate effects of which death took place. The most prominent abnormal nervous phenomenon manifested by this patient, which continued during her whole life, and included most of her multifarious ailments, was excessive general hyper-æsthesia with great mobility of the nervous system.—(*Ed.*)

CHAPTER VIII.

Skin and its Appendages.

SECT. I. SKIN.—The skin consists of two distinct Epidermis. layers: the first, superficial, and composed entirely of cells, is the *epidermis*, or cuticle; the second, beneath this, which has for its basis a dense interlacement of fibres of connecting tissue, containing in its meshes a quantity of nerves, blood-vessels, glands, and masses of adipose cells, is the *derma*, or true skin.

In the epidermis there are three distinct strata. The first, in contact with the true skin, consists of a single layer of cylindrical cells with well marked nuclei, and with their long diameters at right angles to its surface (Pl. XXIII. fig. II. 3; fig. III. 1; fig. IV. 3). It is principally in this layer that the deposit of black pigment is found which causes the dark color of the negro, and in certain regions of the skin in the white, as, for example, the nipple and the scrotum (fig. III. 1). Upon this stratum of cylindrical cells is another, five or six times the thickness of the first, and likewise consisting of nucleated cells (fig. I. 3; fig. II. 2). The deepest cells of this layer are oval, the next in order, round, or regularly polygonal, and the most superficial again oval; but their long diameters have a different direction from those of the deep cells; that is, they are parallel to the cutaneous surface. It is these two united layers which form the

rete mucosum of Malpighi. Finally, the third stratum
or horny layer, very variable in its thickness, is composed entirely of scale-like cells, which overlie each
other regularly (fig. I. 2; fig. II. 1), and which differ
from the cells of the *rete mucosum*, not only in their
form, but also in the absence of their nuclei, and in
their contents, which are more or less opaque, and
coarsely granular. They resist also for a longer time
the action of acetic acid, and caustic potash.

The epidermis, like the other epithelial membranes,
has neither vessels nor nerves, but it does not, on this
account, possess in any less degree true organization
and indubitable vitality.

Derma.   The true skin is naturally divisible into two strata,
which insensibly mingle with each other along their
line of contact. The deep, or reticulated layer is
made up of a loose interlacement of connective and
elastic fibres which, on its internal surface, become
continuous with those which go to form the superficial fascia. It is in this reticulated stratum of the
derma that we find its glands, the hair follicles, and
fat cells, grouped together in little rounded masses.
(Pl. XXIII. fig. I. 8, 10.) It is here also that its
blood-vessels ramify and give off their ultimate
branches, which are distributed to the superficial
layer.

Papilla.   The surface of the superficial or papillary layer of
the true skin is studded with minute projections
known as its *papillæ*. These *papillæ* are not everywhere uniformly distributed, nor do they present
everywhere the same volume; they are most numerous
and largely developed on the extremities of the fin-

gers and toes, and upon the palms of the hands and soles of the feet; some of them, in these localities, are even surmounted by secondary papillæ.

The papillæ, as well as the portion of the derma from which they project, are composed of very delicate fibrous tissue, containing a large number of plasmatic cells (fig. II. 5), and tunnelled by the terminal branches of the vascular and nervous systems. The papillary surface of the true skin is limited by a very delicate structureless membrane ($\frac{1}{5000}$th of a line in thickness), which separates it from the epidermis (fig. II. 4).

The blood-vessels of the skin form two sets: one occupies the deep stratum, and supplies the glands, hair follicles, and pellets of fat; the other, in the shape of a close network, is found spread out in the superficial layer, where it gives off the terminal loops which penetrate the interior of most of the papillæ. *Vessels.*

The nervous filaments of the deep layer, few in number, are destined for the supply of the organs which it contains, whilst those of the papillary layer are very numerous, forming a plexiform network which seems to terminate, after previous sub-divisions, by free extremities. A large number of these ultimate nervous filaments enter the bases of a certain proportion of the papillæ (nervous papillæ), and terminate there either by free extremities, more rarely by forming loops, or lastly by olive-shaped extremities, which constitute the tactile corpuscles already described (Pl. XXIII. fig. II. 6). It will be recollected, also, that the Paccinian corpuscles constitute another mode *Nerves.*

of termination of the cutaneous nerves (Pl. XIV. fig. II).

Lymphatics. The lymphatics of the skin form a very close web in the papillary layer, communicating by larger branches with the subcutaneous vessels of the same system. We possess no positive information as to their mode of origin. M. Küss asserts that they are in direct communication with the deep stratum of the epidermis.

The cutaneous glands have been already described.

Development. According to Bischoff, the skin can be distinguished as a distinct membrane as early as the commencement of the second month of fœtal life. The true skin, as yet consisting entirely of embryonic cells, very soon acquires an increased degree of density, and becomes distinct from the epidermis; later, the cells are metamorphosed, some of them into connective and elastic fibres, and others into vessels, etc. Finally, a certain proportion of them seem to undergo a temporary arrest in their development, and constitute what have been denominated plasmatic cells. As for the epidermis, it is developed by the multiplication and increase in volume of its globular elements. Although the fact has not as yet been demonstrated, it is probable that this multiplication of cells is effected, both in the embryo, and throughout life, by the endogenous generation of new nuclei, and subsequent cleavage of the parent cell.

Preparation. To study its structure, very thin sections both of recent and dried skin, must be made by a razor. Acetic acid renders these sections more transparent, and caustic potash has the same effect, but breaks down their tissue; these reagents, therefore, are useful in bringing out their details of structure. Tactile corpuscles are best found in sections of the skin of the palmar surface of the third phalanges of the fingers, and in the corresponding portions of the toes.

SECT. II. NAILS.—The substance of the nails is Nails. nothing more than a peculiar form of hypertrophy of the epidermis. The deep surface of this horny plate rests directly upon the true skin; its anterior margin is free, whilst its posterior and lateral borders are received into a groove formed by a fold of the true skin (Pl. XXIV. fig. I. II. III.).

On the surface of the derma, in contact with the Matrix. deep surface of the nail, known as its matrix, we find numerous delicate ridges which run parallel with each other and with the axis of the limb, converging at the root of the nail towards a common centre, and usually possessing no papillæ. With these exceptions the structure of the matrix of the nail is the same as that of the true skin elsewhere; its two strata exist as usual; it is to be noticed, however, that its vascular network is less rich towards the root of the nail, causing a dead white surface of the semilunar shape (the *lunula*) partly covered by the margin of the fold of the skin, into which the posterior border of the nail is received.

It consists of both of the layers of which the skin is formed, but the dermal layer is destitute of papillæ (fig. I. 5).

In the nail we find a repetition of the three strata Structure of the of cells which we have seen forming the epidermis. nail. The middle and deep layers are identical with those of the epidermis, and continuous with them without any line of demarcation (fig. II. 2, 7). In the superficial stratum, the distinctive characteristics in which it differs from the corresponding layer of the cuticle are: the persistence of the nuclei of its cells, the greater transparency of their contents, and a firmer cohesion of the cells to each other.

The epidermis of the fold at the root of the nail adheres closely to its surface (fig. II. 10), but as its component cells are not exactly identical with those of the surface of the nail, there results a very distinct line of demarcation between the two otherwise continuous layers (fig. II. 4).

Development. As early as the third month, the groove which receives the root and sides of the nail is apparent. Both the true skin and epiderm become slightly hypertrophied, and by the fifth month, the laminated ridges of the derma have become visible, and the nail is distinguishable from the surrounding cuticle—with which it is continuous.

Growth. The growth of the nail takes place at the expense of the *rete mucosum*, the more superficial cells of which become successively transformed into scales of horn. Its increase in length is effected by the active vegetation of the cells at the bottom of the groove which lodges its root, they becoming continuous with its substance. Whilst it is thus pushed forwards, the deep surface of the nail appropriates a portion of the cells generated from the surface of its matrix, and it thus grows in thickness; but the growth in length is by far the more rapid of the two processes.

In the study of its structure, similar preparations

are required to those of the skin; the use of dilute solution of potassa is required to separate the cells of the horny lamina.

SECT. III. HAIR.—The hairs, like the nails, are composed of epithelium. A hair is a delicate cylinder, generally more or less flattened, variable in its dimensions, and consisting of two distinct portions: the shaft, which projects beyond the surface of the skin, and the root which · is imbedded in a sheath, or follicle, furnished by the skin. This latter terminates by a bulbous enlargement, at the extremity of which is a deep excavation, which is occupied by one of the papillæ of the skin (germ, pulp, papilla of the hair, Pl. XXIV. fig. IV. 3, 7). *Hair.*

The surface of the hair is formed by a single layer of epithelium, called the epidermis of the hair (Pl. XXIV. fig. VII. 1); immediately beneath this is found a material arranged in longitudinal striæ which constitutes almost the entire bulk of the hair, and which is known as its cortical substance (fig. VII. 2); finally, in the centre of the shaft there is generally, but not always, a canal filled by cells of a peculiar shape, which forms its medullary substance (fig. VII. 3). *Structure.*

The *epidermis* is composed of a single layer of scaly cells, presenting an imbricated arrangement, so that their superior margins are free (Pl. XXIV. fig. V.). On treating this layer with acetic acid, or caustic potash, its · cells swell out and become more transparent, so that it can be seen that their contents consist of fine granules, and that their nuclei have disappeared (fig. VI.). The epidermis is found only *Epidermis.*

upon the shaft of the hair, it ceases abruptly at the commencement of its root. Leydig has represented, as the epidermis of the root of the hair, a layer of cylindrical cells, which are arranged perpendicularly to its surface, but their existence is by no means constant.

Cortical substance.

The cortical substance, the color of which varies with the hair, is marked by longitudinal striæ and linear spots, which run in ·the same direction (Pl. XXIV. fig. VII. 2). The elements of which it is composed cohere very closely, but, by the aid of caustic potassa, they can be readily separated and recognised as long fusiform bodies, homogeneous in their structure, without trace of nuclei, and containing, sometimes, pigmentary granules. The dark linear spots seem to be simply cavities filled with air, for they are found in white, as well as in colored hairs (Pl. XXV. fig. I.).

In the root of the hair the cortical substance presents again a different appearance; in the bulb we find regular polygonal cells with clearly defined nuclei, and granular contents sometimes transparent and at others charged with pigment (Pl. XXV. fig. II. 6). A little higher up, these cells, and their nuclei also, become elongated, and the outlines of the cells gradually grow pale and disappear, whilst their nuclei continue to elongate, and remain visible. They finally, however, become pale and disappear also, or, perhaps, they are converted into the fusiform bodies of the cortical substance after their cell walls are absorbed?

Medullary substance.

The medullary canal does not always exist, and,

when present, it varies in shape and length. Thus, sometimes it occupies the whole length of the hair; at others, it ceases at the root; and again, it frequently presents constricted portions, and even entire interruptions (Pl. XXIV. fig. IV. 6, fig. VII. 3 ; Pl. XXV. fig. II. 8). The cells by which it is filled constitute the medullary substance; they contain, usually, a very pale nucleus, and fatty looking granules; according to Kölliker they contain also air bubbles.

The hair follicle, or sheath, which envelops its root, Hair follicle. recalls in its structure that of the skin; in fact we find, on section of its walls, two layers of connecting tissue similar to those of the skin, and two other strata of cells representing the epidermis. This structure suggests the idea, although the mode of development of the hair proves the contrary, that the hair follicle is formed by an indentation, or an involution, of the skin. The following, however, is its structure:

Proceeding from without inwards, we recognise: Structure. 1st, a stratum of very loose connecting tissue, which is continuous with the deep layer of the true skin (Pl. XXV. fig. II. 1); 2d, another stratum, very distinct from the preceding, and similar in structure to the papillary layer of the true skin, with which it is also continuous; internally this layer is limited by a delicate and structureless basement membrane, as in the skin (fig. II. 2, 3); 3d, reposing upon this delicate line is the deep epidermic layer, made up of cells whose shape and mode of stratification resemble those of the *rete mucosum* (fig. II. 4, Pl. XXVII. fig. V. 4); 4th, the internal epidermic layer, which, with its scale-like cells, destitute of nuclei, is exactly similar to the horny surface of the epiderm; it terminates by growing generally thinner in the upper third of the follicle, about where the excretory duct of the sebaceous glands appended to the hair pours out its secretion (Pl. XXIV, fig. IV. 10; Pl. XXV. fig. II. 5); P. XXVII. fig. V. 3). A very small fasciculus of unstriped muscular fibres is attached to the bottom of the hair follicle externally, on the side towards which the hair inclines; it passes obliquely upwards to the surface of the true skin, where it is inserted; this is the muscle, by the contraction of which the hair is made to straighten up at right angles to the surface of the skin, or, as the phrase is, to "stand on end."

Papilla.   The papilla which occupies the excavation at the bottom of the bulb belongs to the true skin, and is more delicate in its structure than its other papillæ. It contains several vessels which form loops in its interior, but its connexion with the nervous system is as yet unknown (Pl. XXV. fig. II. 7).

Development.   The hair and the two epidermic layers of its follicle, are developed from a minute mass, or granulation of the *rete mucosum* which imbeds itself in the surface of the cutis, whilst the two outer strata of the walls of the follicle are derived from the formative cells of the true skin. The hair is developed all in one mass, from the little germ at the bottom of its follicle; the cells composing which, by their various transformations, produce the several elements of which it consists, as well as the two epidermic layers by which the follicle itself is lined.

CHAPTER IX.

Intestinal Mucous Membrane.

THE mucous membrane of the alimentary canal is General struc-
ture. continuous with the skin, and, in its essential consti-tuents, possesses a similar structure; thus, it consists of two layers, one of which corresponds to, and resem-bles closely, the superficial stratum of the true-skin— the mucous membrane proper; the other, the ana-logue of the cuticle, and like it composed of cells, is the epithelium. Nevertheless, the varying shape and disposition of its epithelial cells, and the peculiar modes of distribution of its blood-vessels, as well as its glands, constitute a membrane *sui generis* which requires a description in detail.

A general examination of this membrane shows at once that its appearance and structure are not the same throughout. There is evidently a considerable differ-ence existing between the several segments of the intestinal tube; and in view of this fact, and to faci-litate our examination of the membrane, we shall study its structure successively: 1st, in the mouth; 2d, in the pharynx and œsophagus; 3d, in the stomach; 4th, in the small, and lastly in the large, intestine.

The mucous membrane of the lips, cheeks, palate, Mucous mem-
brane of the
mouth. and gums, resembles exactly the superficial stratum of the true skin. It consists of a layer furnished with papillæ at least as numerous, and of the same shape, as those of the cutis, and presenting a structure, which, though perhaps a little more delicate, contains the same elements. There is a striking analogy also in the distribution of its vessels and nerves, but up to the present time no tactile corpuscles have been observed, except in the papillæ of the lips. On the hard palate and the gums the mucous layer is strongly adherent to the periosteum, with which, in fact, it is continuous; but on the cheeks and lips, it is rein-forced by a delicate fibrous layer, by which it is loosely connected with the subjacent muscles.

Epithelium.   Its epithelial investment presents the same general physiognomy as the epidermis; the two deeper strata, corresponding to the *rete mucosum* of the cuticle, are absolutely identical with it, both in the shape of their cells, and in the order of their superposition. The superficial layer is likewise composed of scaly cells, but differing from those of the skin in the persistence of their nuclei.

Glands.   The glands of this portion of the alimentary mu-cous membrane are all of the clustered variety; they occupy the sub-mucous stratum, sometimes even, as in the cheeks, being imbedded in the muscular layer; on the hard palate and gums they are absent.

Mucous mem-brane of the tongue   The mucous coat of the inferior surface of the tongue is similar to that of the lips, but on its dorsal aspect it is very different, both in its external appear-ance, and in certain structural details.

On the base of the organ its mucous membrane is almost smooth, and sparsely studded with little len-ticular prominences, with a hole in the centre of each, formed by the projection of small masses of subjacent ductless follicles. The rest of its dorsal surface is thickly covered with very prominent papillæ, which from their variety in form, are divided into three sets, viz: *circumvallate, fungiform*, and *conical*, or *filiform*, papillæ.

The papillæ of the first set are found immediately Circumvallate or calleiform papillæ. in front of the base of the tongue, where they are arranged in the shape of the letter V. They have the shape of an inverted cone, with the apex continuous with the membrane below, and its base looking upwards and free, and are moreover surrounded by an elevated circle formed by the mucous membrane, which constitutes a sort of imperfect capsule, in which they are almost concealed (Pl. XXV. fig. III. 1).

The *fungiform* papillæ have the same shape as the Fungiform papillæ. last, but are smaller, and project farther from the surface of the membrane. They are pretty uniformly distributed over the whole surface of the tongue, being somewhat more numerous at its border and tip.

The *filiform* papillæ, whose shape is very well Filiform papillæ. indicated by their name, although found everywhere on the upper surface of the tongue, are more numerous in the neighborhood of its median line than its edges, where they lose their characteristic appearance (Pl. XXV. fig. IV.). Their direction is obliquely upwards and backwards.

Each papilla, whatever may be its shape, is com- Structure. posed of the mucous membrane proper, of the substance of which it is a projection, and covered externally by epithelium. In each of the three varieties, the inner or proper surface of the papilla is studded with secondary papillæ, in the shape of slender processes of variable length (Pl. XXV. fig. III. 2; fig. IV. 2). As for the epithelium, it adapts itself accurately to the subjacent membrane, presenting a free surface which differs in appearance according as it corresponds to the localities occupied by the circumvallate and fungiform papillæ, or to those of the filiform variety; in the first case it is perfectly smooth (fig. III. 3), but in the latter it presents long and delicate filaments, variable in length, and identical in shape with the secondary papillæ, the surface of which they cover (fig. IV. 3, 4). It is this epithelium of the filiform papillæ which in some animals assumes a horny character, and thus constitutes a prehensile organ; in the pike its structure is identical with that of its teeth. The epithelium of the tongue is identical with that of the lips and cheeks, both as regards the form and disposition of its cells.

The filiform papillæ differ from both of the other varieties, not only in their epithelial aspect, but also in their relations to the nervous system; thus the nervous filaments which penetrate them are few in number, and they do not reach their secondary papillæ. The circumvallate and fungiform papillæ, on the contrary, are relatively rich in nerve fibres, and they can be traced readily into their secondary papillæ, where they seem to terminate by free extremities; Kölliker has even demonstrated the presence of tactile corpuscles in the fungiform papillæ of the tip of the tongue.

The blood-vessels which they receive are distributed similarly to those of the other papillæ of the buccal cavity.

The glands of the mucous membrane of the tongue Glands. are of two sorts : clusters of follicles, and ductless follicles. The former, strictly speaking, do not belong to the membrane, for they are imbedded in the surface of the muscular tissue of the organ. They are very numerous at its base, where they form a continuous layer which·extends, on either side, to the pillars of the fauces,.and, in front, encroaches upon its papillary surface. Those which are situated posteriorly upon the edges of the tongue, and upon the under surface of its tip, are also concealed amongst the superficial muscular fibres of the organ, and are connected with its mucous membrane by their excretory ducts only, which pierce it, and open either at the bottom of the *sulci* on its edges, or on either side of the *frœnum.*

The ductless follicles at the base of the tongue Ductless glands. constitute the little lenticular eminences which are found in that region. Upon the summit of each little projection is an orifice, visible to the naked eye, which leads into a flask-shaped *cul-de-sac*, the walls of which are continuous, and identical in structure, with the mucous lining of the organ. The network of vessels which immediately surrounds this orifice is closer and richer than elsewhere in its vicinity (Pl. XXV. fig. V.). The walls of the cavity are reinforced by a dense lamina of connecting tissue, in the substance of which are imbedded about twenty minute spherical bodies, of the same size as the solitary glands of the intestine, and identical with them in structure (*vide ut supra*). These follicles present no trace of an excretory duct, or any visible external opening; the existence of the central orifice on the surface of the mucous covering of the eminence which they form, has led to mistakes on this point, but, as we have already seen, this opens only into a cul-de-sac.

Tonsils. The *tonsils* are composed of an aggregation of ductless follicles identical with those just described; they are therefore compound ductless glands. The excavations which we see upon their free convex surfaces are simple blind cavities, sometimes containing masses of whitish material of a disagreeable odor, and consisting of debris of epithelium in a state of partial fatty degeneration.

The lymphatics of the cavity of the mouth are very numerous, especially those of the mucous membrane of the tongue. They appear to take their origin immediately beneath its epithelial layer, and communicate with the cervical lymphatic glands.

Mucous membrane of the pharynx. The papillæ of the mucous membrane of the pharynx are smaller, and less numerous, than those of the mouth. Its epithelium, in strata, resembles that of the buccal cavity, only it is to be noticed that in the upper portion, or vault, of the pharynx, it is provided with ciliated cells. In the deeper layers of the membrane proper are numerous ductless glands, and clusters of follicles; the first are found only in the upper part of the pharynx, whilst the mucous follicles exist in all parts of the membrane. Its blood-vessels are numerous, and are similarly distributed to those of the walls of the cavity of the mouth. Its lymphatics run into the deeper cervical glands. The nerves are very numerous, and seem to terminate by free extremities.

In the mucous membrane of the *œsophagus* there are a great many conical papillæ, but otherwise its structure is the same as that of the pharynx. It has no ductless glands, and its mucous follicles are not very numerous. Its blood-vessels, by no means so abundant as those of the pharynx, have no peculiarities in regard to their distribution. Its lymphatics communicate with the deep glands of the lower·part of the neck, and with those of the posterior *mediastinum*. It is freely supplied with nerves, but their mode of termination is as yet undetermined.

The mucous membrane of the pharynx and œsophagus is everywhere connected by its attached surface to a thick stratum of muscular tissue; in the pharynx this is formed by its *constrictor* muscles, and in the œsophagus by two layers, of which the fibres of the inner are circular, and the outer, longitudinal. The muscular walls of the pharynx are composed exclusively of striped fibres, but in the œsophagus this is true only of its upper part; below, its muscular fibres are non-striated.*

The mucous membrane of the stomach is softer and

Mucous membrane of œsophagus.

Gastric mucous membrane.

---

* According to Todd and Bowman (*Physiological Anatomy*, Lond. 1856, vol. II. p. 188), striped muscular fibres can be traced as far as the diaphragm, in the muscular coat of the œsophagus; and according to Sharpey and Quain (*Elements of Anatomy*, 5th Ed., Lond., 1848, vol. II. p. 1015), "they have been traced throughout its whole length, and even, it is said (Ficinus), upon the cardiac end of the stomach." This is also stated by Cruveilhier, on the authority of Valentin and Ficinus (Anatomy, 1st Am. ed. New York, 1844, foot note, p. 323.—(*Ed.*)

thicker than that of the œsophagus; it is of a pale
rose color whilst the organ is empty, but becomes
red during digestion. When relaxed, it is thrown
into a great number of wrinkles (*rugæ*), which are
effaced when the stomach is distended. Its surface
is smooth. throughout its whole extent, except near
the cardiac orifice, where there are papillæ similar to
those of the œsophagus (Berres*), and at the pylorus,
where flattened villi are found (Krause†). Its epithelium consists of a single layer of cylindrical cells,
similar to those of the intestine. The reddish tint,
which it derives from the subjacent layers of muscular tissue, contrasts strongly with the whiter color
of the epithelium of the œsophagus; the serrated
line, at the cardiac orifice of the stomach, which
marks the union of these two layers of epithelium, is
very distinct.

Glands.   The surface of the gastric mucous membrane is
pierced by an infinite number of minute holes, from
$\frac{1}{110}$th to $\frac{1}{70}$th of a line in diameter; these are the orifices of the gastric glands (Pl. XXV. fig. VII. 1).
These glands all belong to the tubular variety, but
some of them are single, and others compound. The
former (the glands which secrete gastric juice) occupy
nearly the whole extent of the membrane, and are
the same in form and structure as the follicles of Lieberkuhn already described. As for the compound

* Joseph Berres, Professor of Anatomy in the University of Vienna,
predecessor and preceptor of Hyrtl, the present occupant of the same
chair. Berres died in 1846, leaving an unfinished work on Microscopical Anatomy.—(*Ed.*)

    † O. F. J. Krause, Professor of Anatomy at Hanover.—(*Ed.*)

gastric follicles, one portion of them occupies the
vicinity of the cardiac orifice (those secreting pepsin),
whilst the other (consisting of mucous glands) is
found near the pylorus; both have been already
described (Pl. XXV. fig. VIII.; Pl. XXVI. fig. I.).

The mucous coat of the stomach is rich in blood- Vessels and nerves.
vessels which, first supplying its glands, terminate
nearer its surface in a very regular capillary network,
the largest meshes of which surround their orifices.
Its lymphatic vessels communicate with the little
glands which lie along the greater or lesser curvatures
of the organ.. The mode of distribution and ultimate
termination of the very numerous nervous branches
which it receives from the great sympathetic and
pneumogastric, are not clearly demonstrated.

The mucous membrane of the small intestine, which Small intestine.
in all the essential points of its structure resembles
that of the stomach, differs from it, nevertheless, both
in the appearance of its surface, and in the character
of certain of its glands.   Upon its free surface two
species of prominences are noticeable—*valvulæ conniventes* and *villi*. The former are long semilunar
folds, formed by the plaiting of the membrane upon
itself; their direction is perpendicular to the axis of
the canal, and each occupies the half or two-thirds of
its circumference; they slope. off to a point at either
extremity, and are. connected to each other by little
oblique folds. They are very large and numerous in
the *duodenum*, where they overlap each other like
shingles on the roof of a house, and in such a manner
that their free edges look downwards. As we trace
them downwards, following the surface of the bowel,

they gradually and regularly diminish both in number and in size, until, in the lower part of the ileum, they are recognisable only by a faint thickened line.

The *villi* of the small intestine are minute processes, analogous to the papillæ of the tongue, but more delicate in their proportions. The best idea of their shape, number, and mode of arrangement, is to be got by placing a piece of the mucous membrane under water, and examining it with a magnifying glass, or a microscope with a low power. They are seen to occupy the whole extent of the surface of the membrane, and to be more numerous in the *duodenum* and *jejunum*, than in the *ileum ;* they are seen also to assume two principal forms—the flat, or valvular, and the conical. The flat *villi* are principally found in the upper portion of the small intestine; they are simple and solitary, or, by running into each other, become compound, in which case they resemble minute *valvulæ conniventes* (Pl. XXVI. fig. II. 2, 3). The conical *villi* are found everywhere throughout the small intestine, but they exist in larger proportion in the *ileum* (fig. III. 1). In some instances, instead of terminating in a point, their apices are slightly bulbous (Pl. XXVI. fig. XII.); their average height is from one-fourth to one-half a line, and their diameter from one-sixteenth to one-fourth of a line.

Structure of the villi. Whatever may be the form of a *villus* its structure is always the same, and, in examining it from its surface inwards we find, first: a single lamina of epithelium which, on a perfectly fresh specimen which has not been roughly handled, presents the appearance of a mosaic, upon the surface of which an unbroken layer

of amorphous material has been applied (Pl. XXVI.
fig. IV. 1, 2, 3; fig. V.). By breaking up this layer
of epithelium we recognise the elements of which it
is composed, viz. conical cells, the summits of which
rest upon the surface of mucous membrane, whilst
their bases are directed outwards and free, or rather
covered by the amorphous substance already mentioned, which adheres to them very closely (fig. VI.).
It is only in perfectly fresh and recent specimens that
the cells are found covered by this amorphous coating; it disappears entirely in from twelve to twenty-
four hours after death (Pl. I. fig. VI.). The contents
of a cell consist of fine granules, and its nucleus, which
is oval in shape, is usually nearer to its apex than its
base; their mean length is $\frac{1}{70}$th of a line, their breadth
$\frac{1}{130}$rd of a line, and the diameter of their nuclei $\frac{1}{315}$th
of a line. The epithelium which covers the mucous
membrane in the intervals between the villi has also
the appearance and structure of that first described.
Immediately beneath the layer of epithelium is the
surface of the naked villus, or papilla, and this is
formed by a structureless basement membrane similar
to that already noticed upon the surface of the papillary layer of the true skin.

Beneath this simple membrane is a close network of
capillary vessels forming a sort of hollow bulb which
encloses the remainder of the villus (Pl. XXVI. fig.
VIII.); this capillary plexus seems to communicate
more freely with the venous than with the arterial system, for it is much easier to fill it with fine injection
from the *vena portæ* than from the abdominal aorta.
In the centre of the villus is a large hollow canal ($\frac{1}{135}$th

to $\frac{1}{150}$d of a line in diameter) ending by a blind and somewhat bulbous extremity; this is the origin of a lacteal vessel (Pl. XXVII. fig. VI. 2). The situation of this solitary vessel in the centre of the villus, compared with the position, on its surface, of the capillary plexus, so rich in blood-vessels, should teach us that the activity of its function as an absorbent is subordinate to that of the elements by which it is surrounded. The remainder of the villus consists of a material faintly fibrillated rather than amorphous, which encloses a number of oval nuclei, the meaning of which is unknown (Pl. XXVI. fig. VII. 3). Around the outer surface of the lacteal, non-striated fibres of muscular tissue are sometimes found, running parallel with the axis of the villus. We know nothing of the connexion of the nervous system with the villi.

The glandular apparatus of the small intestine is composed of clusters of follicles, tubular glands, and ductless follicles.

Glands of Brunner.

Brunner's glands are clusters of follicles, situated deeply in the mucous membrane, or rather in the submucous layer of connecting tissue. They are little yellowish-white granules, averaging one-half a line in diameter, and precisely similar in structure to the salivary glands; they are provided with a single layer of polygonal epithelium, and they secrete an alkaline fluid in which no organic element is discoverable (Pl. XXVI. fig. IX.). These glands are only found in the *duodenum*.

Glands of Lieberkuhn.

The tubular glands, or *follicles of Lieberkuhn,* placed side by side like quills in a bundle, constitute a secretory apparatus which occupies the whole

extent of the small intestine.   On examining the surface of a piece of mucous membrane under water,
with a low magnifying power, we recognise a great
number of minute holes, which are nothing more than
the orifices of these glands (Pl. XXVI. fig. II. 4; fig.
III. 2; fig. XII. 3).   Their structure has been already
studied.

The ductless follicles, with the structure of which   Ductless glands.
we are already familiar, are either solitary or aggregated in groups, and in either case are imbedded in
the sub-mucous tissue.   As solitary glands, they are
scattered throughout the whole extent of the mucous
coat of the *jejunum* and *ileum*.   It is only in excep-
tional cases that they are found in the *duodenum*.
When collected in groups they constitute the essential element of the *patches of Peyer*.   These latter,
very variable in number, are usually seated in the
ileum and lower half of the *jejunum ;* sometimes, but
very rarely, they are found higher up, and even in
the *duodenum*.   They are oval in shape, with their
long diameters parallel with the axis of the intestinal
canal, and are seated opposite to the attachment of
the mesentery.   The surface of a patch of Peyer is
studded with villi, and also with the minute orifices
of Lieberkuhn's follicles, as elsewhere on the surface
of the intestinal mucous membrane ; but besides these
it presents a number of larger depressions (one-half a
line in measurement), at the bottom of each of which
is a little prominence which corresponds to the position of a ductless gland.   These are perfectly blind
depressions, situated just over the glands, and this
relation between the two has given rise to the false

impression that these glands possess outlets. The solitary glands, instead of underlying a depression, on the contrary cause a slight projection of the membrane which covers them, which, with this exception, presents the same appearance as elsewhere (Pl. XXVI. fig. XII. 1).

Mucous membrane of large intestine. The mucous membrane of the large intestine is smooth and destitute of villi, and in this respect resembles that of the stomach. Its glandular apparatus comprises follicles of Lieberkuhn and solitary glands, identical with those of the small intestine, only it is to be noticed that its solitary glands correspond in situation always with a depression on the surface of the membrane, as we have seen in the patches of Peyer (Pl. XXVI. fig. XV.). Its vessels present the same appearance and distribution as those of the stomach; in relation to its nerves nothing precisely is known.

The intestinal mucous membrane of the abdomen is connected to the muscular coat of the canal by a rather lax stratum of connecting tissue (sub-mucous layer, fibrous coat, nervous coat) in which are imbedded the glands of Brunner, the ductless follicles—solitary and aggregated, and the sub-mucous network of blood-vessels.

The development of the glands of the intestine takes place by means of minute granulations which spring from its epithelial layer, and in this respect presents a close analogy with the mode of development of the glands of the skin.

The mode in which the epithelium of the intestine is reproduced is unknown at the present time; it is probably accomplished by endogenous vegetation and subsequent cleavage of its cells; the presence of nuclei in a large proportion of these elements renders it probable that the process is thus effected.

CHAPTER X.

Organs of Sense.

SECT. I. THE EYE.—The apparatus of vision comprises the globe of the eye, or the organ of sight properly so called ; its organ of protection, the eyelids ; and its motor and lachrymal apparatus. As the two latter present no features of special histological interest, we shall omit their consideration.

Eyelids   The several layers of tissue which enter into the composition of the eyelids, proceeding from without inwards, are : the skin, the orbicular muscle, the fibrous stratum, and finally, the mucous membrane.

The skin is exceedingly delicate, but otherwise it presents the same structure as elsewhere. Situated deeply in its substance, and near the free edges of the eyelids, we find the hair-follicles of the *cilia* or eyelashes, surrounded by their sebaceous glands (Pl. XXVII. fig. V.). The orbicular muscle of the eyelids belongs to the class of striped muscles (Pl. XXVII. fig. II.). The fibrous stratum, very thin at the bases of the eyelids, becomes more dense towards their free edges, where it forms the tarsal cartilages. These are composed of connecting tissue very much condensed, and studded with plasmatic cells. We search in vain in this stratum of fibrous tissue for cartilage cells, or at least, they are so rarely encountered that they cannot be regarded as constituting one of its normal elements. We must give up the idea therefore that these dense laminæ consist of fibro-cartilage, for they have only its appearance, without possessing its structure. On their deep surfaces there are from twenty to thirty parallel grooves, which accommodate the *Meibomian glands* (Pl. XXVII. fig. III.).

The mucous membrane, or conjunctiva, consists of quite a dense layer of connecting tissue, the ocular surface of which is studded with numerous papillæ analogous to those of the skin ; its epithelium, which is stratified, resembles that of the skin and mucous membrane of the mouth (Pl. XXVIII. fig. II. 5). This membrane, as is well known, is reflected from the eyelids upon the globe of the eye, to which it becomes intimately attached; tracing it here to the circumference of the cornea, it is to be remarked that its deep layer becomes gradually thinner, and ceases suddenly by becoming inserted into the amorphous border which surrounds the anterior margin of the cornea (Pl. XXVIII. fig. II. 4), whilst its epithelial layer continues its course and covers the whole anterior surface of the cornea (fig. I. 7 ; fig. II. 5). The vessels and nerves of the cornea present no peculiarities worthy of note.

Globe of the Eye.—The first proper coat of the eye- Sclerotica. ball is formed posteriorly by the *sclerotica*, and in front, by the *cornea*. The sclerotica is an exceedingly dense membrane, thicker anteriorly and posteriorly than around the centre of the eye-ball, and consisting of a close tissue of connective and elastic fibres. Anteriorly this membrane adheres very intimately to the conjunctiva, which is distinguishable from it by the greater looseness of its texture, and by the larger number of plasmatic cells which it contains (Pl. XXVIII. fig. II.). Its deep surface is closely attached to the *choroid membrane* only at its anterior limits, where it gives insertion to the *ciliary muscle* (fig. IV. 1). The *canal* of *Schlemm* is also found along the line which limits the sclerotica in front, forming a tunnel near the deep surface of the membrane (fig. IV. 2).

Cornea.   The *cornea* is composed of an amorphous fundamental substance, which contains a great quantity of plasmatic cells (fig. I. 2 ; fig. III. 2). These are disposed very regularly in concentric lines running parallel with the two surfaces of the cornea. When very dilute acetic acid is applied to prepared specimens of the cornea, the stellate shape of these cells, and the numerous anastomoses between their prolongations, are rendered perfectly visible, and their appearance recalls vividly the structure of bone. But if the acid should be too much concentrated the prolongations of the cells become pale, and their bodies alone remain visible (Pl. II. fig. V.). The front surface of the cornea is limited by a thin edge which, in a section, forms an amorphous border from $\frac{1}{300}$th to $\frac{1}{100}$th of a line in thickness, the anterior margin being in contact with the epithelium of the conjunctiva (Pl. XXVIII. fig. II. 3). Its posterior surface also forms an amorphous border, of the same thickness as the latter; it is continuous, by means of fibrous tissue, with the anterior margin of the sclerotica and the ciliary muscle (fig. III. 4 ; fig. IV.).

It is invested by a single layer of pavement cells, which is reflected upon the anterior surface of the iris, and ceases at the margin of the pupil (membrane of Demoins, of Descemet, fig. III. 4, 5).

There is no distinct line of demarcation, as inspection by the unassisted eye would lead us to suppose, between the cornea and sclerotica. The two membranes blend insensibly with each other (fig. I. 3). The fibres of the sclerotica become rarified as they approximate the corneal margin, and they can be clearly seen to be continuous with the branches of its plasmatic cells (fig. II. 1, 2).

The sclerotica has but few vessels, and hardly any Vessels and nerves. There are no blood-vessels in the cornea; nerves. its network of plasmatic cells affords ample circulation for the nutritive fluid. Its nerves are very numerous, and form a rich web of filaments which are found chiefly near its anterior surface (Kölliker); according to some authorities they terminate by free extremities.

The second tunic of the eye-ball is formed poste- Choroid riorly by the choroid, and in front by the iris. The choroid coat lines the internal surface of the sclerotica, very accurately, and is continuous in part with the iris, without any line of demarcation (Pl. XXVIII. fig. I. 11, 12). It is loosely connected by its external surface to the sclerotica, by means of the ciliary vessels and nerves, and an occasional very delicate fasciculus of fibrous tissue; its brownish-black color is explained by the presence of a layer of irregularly branching cells filled with pigment granules (Pl. II. fig. II. 1, 2, 3). Its internal surface is likewise covered by a layer of pigment cells, but these are regular polygons, more numerous, and containing a larger quantity of pigment, than those last mentioned (Pl. II. fig. I.); this layer has no connexion with the *retina*, with which it is in contact. Between these two layers of pigment cells is the proper substance of the choroid membrane.

Ciliary muscles. The greyish thickened ring of tissue which forms the limit, anteriorly, of the choroid, and by which it is firmly united to the sclerotica, is muscular in its nature, and constitutes the principal bulk, or body, of the ciliary muscle; it is also described as the ciliary circle or ring, ciliary ligament, and ciliary ganglion. On its surface it appears to consist of non-striated muscular fibres, parallel in their direction with the antero-posterior axis of the globe of the eye. Anteriorly it grows thin ($\frac{1}{3}$d of a line in thickness), and is inserted, at first, into the inferior wall of the canal of Schlemm, and, a little farther on, into the posterior extremity of a fasciculus of fibres which is continuous with the amorphous border of the cornea, and adherent also to the circumference of the iris, called the *pectiniform ligament* (Pl. XXVIII. fig. IV. 3). The deeper portions of the ciliary muscle, which are in relation with the ciliary processes, consist of interlaced muscular fibres; at least this is the inference to be drawn from the variable aspect presented by the muscular nuclei of the part in a section (fig. IV. 5, 6). Posteriorly, the muscle gives off longitudinal fasciculi which extend as far as the middle of the choroid, and anteriorly, it presents similar fasciculi of fibres which penetrate the iris, and converge towards the pupil.

The ciliary vessels which traverse the muscle become intimately amalgamated, so to speak, with its substance, and thus constitute an erectile apparatus (Rouget). The posterior half of the choroid is composed of vessels united together by very delicate connecting tissue, which contains some plasmatic cells.

Each of the surfaces of the iris is covered by a simple layer of epithelium. That upon its anterior aspect has been already examined (fig. III. 5) ; the epithelium upon its posterior surface (*uvea*) is composed of polygonal pigment cells similar to those of the choroid. In addition to the blood-vessels contained in the iris, we find also, in the substance of this membrane, a ring of muscular fibres surrounding the pupil and connected by its circumference with the converging fibres of the ciliary muscle, and finally a web of connecting tissue full of plasmatic cells. In most eyes, but especially in those of dark color, the majority of these cells contain pigment granules. Iris.

The nerves of the choroid coat and iris (ciliary nerves) are very numerous, and seem intended for the supply of the muscular apparatus belonging to these membranes. Nerves.

The *retina*, which constitutes the third coat of the eye-ball, is coextensive with the choroid coat, beneath which it lies. At the entrance of the optic nerve it is thicker than elsewhere ; in fact a slight prominence is perceptible at this point, which has been designated as the papilla of the retina. At the posterior extremity of the antero-posterior axis of the globe, and consequently to the outer side of the pupil, there is, upon the surface of the retina, an elongated depres- Retina.

sion of a light yellowish color—the yellow spot of Sœmmering. As we trace it forwards, the retina grows thinner, and its anterior border corresponds with the outer circumference of the iris; its nervous matter, however, can be traced no farther than the commencement of the ciliary processes, where it terminates abruptly, by a serrated margin (*ora serrata*).

The researches of H. Müller* demonstrate the structure of the retina to be as follows :

Its outer layer is made up of little "club-shaped rods" arranged closely side by side so as to form an uninterrupted lamina—the *membrana Jacobi*. Their direction is perpendicular to the surface of the retina, and their shape, as their name indicates, is cylindrical; but some of them enlarge at their outer extremities, so as to assume a conical form. These latter are fewer in number than the club-shaped bodies first mentioned, and are quite uniform in their distribution, with this exception, that they alone constitute the whole thickness of Jacob's membrane where it passes over the yellow spot of Sœmmering. The internal extremities of these club-shaped bodies taper off, and each one of them becomes continuous with a fibre (fibre of Müller), which traverses the whole thickness of the retina. In its course this fibre presents three distinct enlargements : the first is situated at its outer extremity, just where it is joined by the corresponding club-shaped body; the second at its middle, and the third at its internal extremity. The first corresponds to the *external granular layer*, which

* Professor, or Lecturer, on Anatomy in the University of Wurzburg, Bavaria.—(*Ed.*)

is made up entirely by these external enlargements or granules, the second to the *internal granular layer*, and the third series of enlargements is intimately mingled with the fibres of the optic nerve, and with them constitutes the *fibrous stratum* of the retina. The external and middle enlargement of the fibres of Müller are composed of cells; the internal enlargement consists of a homogeneous mass, with a depression upon its inner surface, by which it rests upon a very delicate structureless lamella ($\frac{1}{10000}$th of a line in thickness) which constitutes the limitary membrane of the retina within.

On the internal surface of the *internal granular layer*, and consequently between it and the *fibrous* layer, is a stratum composed of multipolar or caudate nerve cells (nervous layer); and the fibrous layer, as already stated, consists of an expansion of the fibres of the optic nerve.[*] These latter, as they run forwards towards the anterior border of the retina, curve outwards to unite themselves with the prolongations of the nerve-cells, which, in their turn, send an anastomosing fibre to each of the middle enlargements on the fibres of Müller. In the centre of the yellow spot of Sœmmering we find nothing but nerve-cells within the *membrana Jacobi*.

In reviewing what has just been said in relation to the arrangement of the diverse elements composing the retina, it is obvious that this membrane presents

---

[*] The superposition of a layer of nerve cells upon a layer of nerve fibres recalls the structure of the convolutions of the cerebrum and cerebellum. The capillary layer of the retina formed by the branches of the *arteria centralis retinœ* occupies the stratum of nerve cells.—(*Ed.*)

a series of distinct strata, which, proceeding from without inwards, are as follows : 1st, the *membrana Jacobi*, composed of club-shaped bodies; 2d, the *external granular layer*, corresponding with the external enlargements upon the fibres of Müller; 3d, the *internal granular layer*, corresponding to the middle enlargements of the same fibres; 4th, the nervous layer, consisting of nerve cells; 5th, the fibrous layer formed by the fibres of the optic nerve; 6th, and last, the limitary membrane.*

Arteria centralis retinæ.

The central artery of the retina traverses the papilla of the optic nerve, and is distributed in the more internal layers of the retina. It forms a vascular circle around the "yellow spot," and terminates, by a second circular plexus, at the *ora serrata*. The vein has the same distribution as the artery.

Vitreous humor.

Interior of the eye.— Vitreous humor.—The vitreous humor of the eye occupies the cavity of the retina, and is adherent to that membrane only in the interval between the *ora serrata* and its anterior margin ; in front it presents a cup-shaped depression which receives the *crystalline lens*. Its envelope, the hyaloid membrane, is very delicate, structureless, and transparent; interiorly it is somewhat thicker than elsewhere, and gives off two laminæ, which invest, respectively, the anterior and posterior surfaces of

---

* To obviate any obscurity in the text, arising from the very compendious and precise style of the author, who always assumes that his reader is a good descriptive anatomist and possesses a fair knowledge of general structure, it would be well for the student to refer to the admirable work of TODD and BOWMAN, " *The Physiological Anatomy and Physiology of Man*," where fuller details, up to the period of its publication, will be found. Vide Vol. II., p. 27. London, 1856.—(*Ed.*)

the crystalline lens, enclosing, at their angle of separation, the canal of Petit, which surrounds the peripheral border of the lens. The vitreous humor,
which is adherent to the internal face of this membrane, is an amorphous hyaline substance which, in
the fœtus, contains oval nuclei and stellate cells; but
in the adult these cellular elements are no longer
visible, and nothing remains beyond an amorphous
jelly-like mass.

The vitreous humor contains neither nerves nor
vessels. During fœtal life its antero-posterior axis is
occupied by a tubular canal, containing a delicate
branch of the *arteria centralis retinæ* sent forward to
the capsule of the lens; after birth this canal becomes
obliterated.

The *crystalline lens* is a solid body, surrounded by
a membranous envelope. This containing membrane,
or capsule, of the lens is highly transparent and
entirely destitute of structure, resembling a delicate
lamina of the purest glass. It possesses great elasticity, but is readily torn; on the anterior surface of
the lens its thickness is $\frac{1}{175}$th of a line, and posteriorly
but $\frac{1}{750}$th of a line. It resists perfectly the action of
boiling water, a solution of potassa, and the acids.
Its external surface is continuous posteriorly with the
hyaloid membrane of the vitreous humor; anteriorly
it is free; its internal surface is lined by a layer of
exceedingly delicate polygonal cells, which, liquefy-
ing shortly after death, form the liquid of Morgagni.*

Crystalline lens.

---

* John Baptist Morgagni, the celebrated professor of anatomy at the
University of Bologna in Italy, and the preceptor of Scarpa, was born in
1682, and died in 1771.—(*Ed.*)

This liquid has been supposed by some to be derived from a different source, viz. from a layer of globules underlying the capsular epithelium, and distinguishable from it by their spherical shape, and by the absence of nuclei in their contents (Warlomont).*

Beneath the epithelium of the capsule, and the globules of Morgagni, is the proper substance of the crystalline lens, which consists of a central portion or nucleus, and a peripheral or cortical portion. The central nucleus is a little star-shaped mass, made up entirely of very minute granules. The cortical portion of the lens is composed of concentric lamellæ, and each lamella, of hexagonal prisms in close apposition and flattened antero-posteriorly. These prismatic bodies are more numerous and delicate in their proportions as we trace them more deeply into the substance of the lens, and they adhere to each other more closely by their edges than by their faces, which explains, on one hand, why the substance of the lens increases in density from its surface towards its centre, and on the other, why it disintegrates more readily into laminæ than into fibres. They consist of a very delicate and structureless containing membrane, with contents of a semi-fluid consistence, equally destitute of structure, and albuminous in their nature; their edges are serrated, and the interdigitation of these serrations adds to the solidity of their union.

Each prism, taking its origin from one of the prolongations of the central nucleus of the lens, passes

* Dr. E. Warlomont, chief editor of the ' Annales d'Oculistique,' published at Brussels, Belgium.—(Ed.)

outwards towards its border which it doubles, and returns upon its opposite surface, where it terminates; its two faces do not give the same measurement in length, which is explained by the fact that its extremities are both beveled, and in opposite directions.

Neither the crystalline lens nor its capsule possesses vessels or nerves, at least in the adult. During fœtal life, the branch of the central artery of the retina, which traverses the tubular canal in the vitreous humor, furnishes branches which encircle the capsule, and afterwards lose themselves in the pupillary membrane, where they anastomose with the ciliary arteries.

The histological development of the eye follows Development. the universal law; all of its component parts take their origin from embryonic cells which take on determinate metamorphoses in order to form each of its individual tissues. Each of the prismatic fibres of the crystalline lens is the result, apparently, of the elongation of a single cell, and not of the fusion together of an indefinite number of cells.

SECT. II. THE EAR.—The skeleton of the external External ear. ear is osseous in the deep portion of the *meatus*, but elsewhere it is fibro-cartilaginous. The integument by which it is invested contains glands of different kinds in its several regions. In the *concha* we find a great many sebaceous glands; we encounter these organs again in the external *meatus*, but here they are in company with ceruminous glands; finally, sudoriparous glands are found everywhere, but principally upon the internal surface of the *concha*.

There is nothing especially worthy of notice in the

mode of distribution of the vessels and nerves of the external ear.

Middle ear. The mucous membrane of the middle ear is very thin; in the Eustachian tube only it is somewhat thicker. From the bony walls of the cavity it is reflected upon the inner surface of the *membrana tympani* to which it is closely adherent, and upon the muscles and *ossicula*, for which it forms a periosteal investment. Its epithelium is everywhere ciliated, except upon the internal surface of the *membrana tympani*, where, according to Kölliker, it forms a simple tessellated layer.

The *membrana tympani* consists of a fibrous expansion made up of radiating and circular fasciculi, and inserted by its circumference into the groove of the temporal bone like a watch crystal into its case; we are familiar with the relation of its internal surface to the mucous lining of the tympanum; its external surface receives a layer of epidermis from the walls of the *meatus auditorius externus*.

The blood-vessels of the middle ear are numerous; they form a rich network in the substance of its mucous membrane, and in the *membrana tympani*. The lymphatic vessels probably accompany its arteries and veins.

The nerves of the middle ear come from the fifth, seventh, and ninth* pairs of cranial nerves; their mode of termination is unknown; Kölliker describes masses of ganglionic cells as existing in the substance of the tympanic nerve.

Internal ear. The bony walls of the semicircular canals, vesti-

* According to the classification of Sœmmering.

bula, and cochlea, are covered by a layer of connecting tissue, with pavement epithelium upon its surface.

The walls of the membranous labyrinth consist of a lamina of extremely delicate connecting tissue, the fibrillated character of which is with difficulty recognizable, but it contains a large quantity of oval (fibroplastic) nuclei, and is covered by a very thin amorphous layer, on the internal surface of which we find a stratum of pavement epithelium.

The white specks which are observed upon the inner surfaces of the *sacculus communis* and *sacculus proprius* (otoliths, otoconites) are composed of calcareous granules, which sometimes present a crystalline aspect. Both outside and inside of the tubes and cavities forming the membranous labyrinth, is a pellucid fluid (perilymph, endolymph), the chemical nature of which is not as yet clearly determined.

The nerves which reach the *ampullæ* of the semicircular canals and the *sacculi* appear to terminate by free extremities, after having undergone frequent division and subdivision; beyond these localities it has been found impossible to trace them.

The vessels form a close network which principally occupies the fibrous coat of the semicircular canals, and the two vestibular sacculi.

The labors of Corti* and Kölliker tend to prove that the nervous fibres of the *cochlea* terminate by free extremities in the substance of the membranous portion of the *lamina spiralis*. These authors have also demonstrated that there are cellular enlarge-

* Corti wrote on the structure of the retina in Müller's Archiv, 1850, p. 274.—(*Ed.*)

ments in the course of these fibres, analogous to those of the nerve fibres of the retina.

The capillary network formed by the blood-vessels of the *cochlea* is equally rich with that of the vestibule and semicircular canals.

Olfactory mucous membrane.

SECT. III. OLFACTORY MUCOUS MEMBRANE.—The mucous lining of the nasal cavities is thick, soft, tomentose, and reddish in color, especially in its inferior two-thirds. Its texture is composed of interwoven fibres—connective and elastic—but the proportion of the latter is small; it contains likewise an abundance of plasmatic cells. Its deeper surface is intimately adherent to the periosteum, and contains a great number of mucous follicles, in clusters. Its free surface is covered by stratified epithelium, and its superficial layer of conical cells is provided with *cilia*. Its vessels are exceedingly numerous, and form a web of unusual thickness. The nervous filaments furnished by the fifth pair are distributed throughout its whole extent, and present nothing worthy of note, but those which come from the olfactory nerve are distributed only to the mucous membrane covering the superior turbinated bone, and the upper third of the septum between the nostrils; moreover, they apparently end by free extremities. According to some authors they present globular enlargements at their extremities, similar to those of the retina.

THE END.

EXPLANATION OF THE PLATES.

PLATE I.

VARIOUS FORMS OF CELLS.

FIG. I.* **Blood of the adult.**—1, Red globules, front view; 2, *same* in profile; 3, *same* altered; 4, white globule.

FIG. II. **Epithelial cells of the bladder.**—1, Cell with granular contents; 2, its nucleus, also containing granules, one of which (3) larger than the rest forms the nucleolus; 4, cell with two nuclei; 5, group of cells retaining their original relation to each other.

FIG. III. **Hepatic cells.**—They are polygonal in shape, and scattered amongst their granular contents free fat is to be seen in the form of brilliant little pearl-like globules.

FIG. IV. **Epidermic cells.**—1, Cells of the *rete mucosum ;* 2, cells of the middle stratum; 3, cells of the surface, in the shape of granular scales without nuclei; 4, other cells from the superficial layer, detached and swelled by very dilute acid.

FIG. V. **Adipose cells** from beneath the integument. Their fluid contents are so transparent that the outlines of the cells alone are visible.

FIG. VI. **Epithelial cells from the small intestine,** examined thirty hours after death.

FIG. VII. **Epithelial cells of the trachea.**—1, Body of

---

* All of the figures contained in the following plates, unless otherwise stated, were drawn from preparations taken from the adult male subject, and magnified 400 diameters by a Nachet's microscope.

a ciliated cell with its nucleus; 2, outline of the layer of amorphous material at its base; 3, cilia; 4, deeper cells of the same layer; 5, normal relation of these cells. The ciliated cells constitute the free surface of the epithelial membrane.

————

PLATE II.

CELLS, *continued;* CONNECTING TISSUE.

FIG. I. **Pigment cells** from the deep surface of the choroid. They are very regular polygons filled with granules of pigment, except in the centre, where the bright spot corresponds with the nucleus.

FIG. II. **Branching pigment cells** from the outer surface of the choroid; 1, cell; 2, nucleus; 3, anastomosing branch; 4, nucleus of oval or fusiform cells, scattered amongst very pale connective fibres.

FIG. III. **Fusiform cells** (fibro-plastic).

FIG. IV. **Adipose cells** containing acicular crystals of margarine in tufts, or solitary.

FIG. V. **Plasmatic cells, or nuclei** of the cornea. Their anastomoses are perceptible.

FIG. VI. **Cells from a cancer** of the heart, in which endogenous multiplication of the nuclei, by cleavage, is seen.

FIG. VII. **Another type of cancer cell**, showing process of endogenous multiplication of nuclei by cleavage.—1, Nucleus in process of cleavage; 2, separate nuclei.

FIG. VIII. **Multiplication of cells.**—1, Process of endogenous formation as observed in fœtal marrow; 2, multiplication by cleavage in the cartilage cell.

FIG. IX. **Superficial fascia** of the forearm. It is composed of fasciculi of connective fibres: 1, wavy and crossing in all directions so as to form an interlacement varying in density. Amongst these fasciculi a number of elastic fibres (2) are to be seen.

PLATE III.

CONNECTING TISSUE, *continued*.

FIG. I. **Connective fibres** in wavy and parallel bundles. On the right of the preparation they have been slightly teased out (*tendo Achillis*).

FIG. II. **Longitudinal section of tendon** (*tendo Achillis*), treated by acetic acid. The fasciculi of connective fibres have grown pale and disappeared. 1, Plasmatic cells in longitudinal rows between the fasciculi of fibres; 2, anastomoses between them (from the fœtus).

FIG. III. **Longitudinal section of tendon** (*Peronæus longus*).—1, Bundles of connective fibres; 2, plasmatic cells.

FIG. IV. **Transverse section of the tendon of the Peronæus longus.**—1, Granular basis indicating the section of the connective fibres; 2, irregular wavy lines marking the intervals between the fasciculi; 3, plasmatic cells.

FIG. V. **Elastic fibres** from a yellow ligament.—1, The fibres in their normal relation; 2, separate fibres.

PLATE IV.

CONNECTING TISSUE, *continued*.

FIG. I. **Longitudinal section of the superior extremity of the tendo Achillis** (of an old man).—1, Fasciculus of connective fibres; 2, plasmatic cells in parallel rows.

FIG. II. **Similar section of inferior extremity** of same tendon. 1, Connective fibres, slightly wavy; 2, cartilage cells which have taken their origin from plasmatic cells.

The following figures show the different modes of development of connective fibres.

FIG. III.—1, **Embryonic cell**; 2, same cell elongated, its contents already divided into fibrillæ; 3, two cells united at their extremities, about to form a bundle of connective fibres.

Fig. IV. **Fibrous tumor of the dura-mater.**—1, Free fusiform cells; 2, fasciculi of same cells united at their extremities ; 3, fasciculi of fibres formed by the elongation of the same cells and the disappearance of their nuclei. In this case each row of cells forms but *one solitary fibre.*

Fig. V. **Fibrous tumor of the uterus** in which the formation of a fibre by metamorphosis of a nucleus can be traced. 1, Finely granular substance ; 2, nuclei.

Fig. VI. **Another portion of same tumor.** The nuclei somewhat elongated in shape already show a disposition to assume the form of fibres.

Fig. VII. **Same tumor.** The nuclei are still more elongated ; at some points they can be seen with their extremities united together so as to form fibres.

PLATE V.

CARTILAGE AND BONE.

Fig. I. **Section involving the centre of a costal cartilage.** 1, Fundamental substance, slightly granular and transparent; 2, cartilaginous capsule; 3, primordial cell, or utriculus; 4, nucleus—made up of fatty granules ; 5, capsule containing four cells, two of which have no nuclei.

Fig. II. **Costal cartilage with its perichondrium,** taken from a subject eighteen years of age. 1, Perichondrium formed by a dense interlacement of connective and elastic fibres, and studded with plasmatic cells. 2, There is no clear and distinct line of demarcation between the deepest portion of the perichondrium and the substance of the cartilage ; it is also almost impossible to make out a distinct difference in the character of the superficial cells of the cartilage and the plasmatic cells of the deepest layer of the perichondrium.

Fig. III. **Fibro-cartilage from the ear.** 1, Fibrous fundamental substance or basis; 2, capsule inclosing these cells.

Fig. IV. **Transverse section of the ulna.** In the midst

of the amorphous fundamental substance of the line are to be seen: 1, the stellate or branching bone-cells (lacunæ); their branches or prolongations (2) in the shape of canaliculi, anastomosing with each other so as to form a network by which a communication is established between the corpuscles themselves, and also with the Haversian canals 3, or with the interior cavities of the bone.

Fɪɢ. V. **Same section** seen with a magnifying power of eighty diameters. The bone-corpuscles (lacunæ), in the shape of minute elongated black spots, are seen to be grouped in concentric circles around the Haversian canals (1).

PLATE VI.

BONE, *continued.*

Fɪɢ. I. **Longitudinal section of the shaft of the femur** (80 diameters). 1, Longitudinal Haversian canals; 2, transverse anastomotic canal; 3, confluence of several canals.

Fɪɢ. II. **Longitudinal section of the condyles of the femur** (in a newly born infant. Magnifying power of 180 diameters). 1, Line of junction of the cartilage with the bone. Above this line the cells of the cartilage are seen grouped in parallel rows. Their nuclei (2) deeply shaded and presenting jagged edges. Below this same line the cartilage is seen, infiltrated with earthy salts and in process of ossification.

Fɪɢ. III. **Section of cartilage** taken from same femur $\frac{1}{13}$th of an inch beyond the newly ossified portion. 1, Fundamental substance, entirely transparent; 2, limit of the capsule; 3, limit of the cell; 4, nucleus assuming a branching character.

Fɪɢ. IV. **Formation of marrow and of medullary cavities** in newly ossified bone (from same femur). 1, Fundamental substance infiltrated at certain points with free fat 2; 3, capsule of cartilage cells; 4, a parent cell full of young cells; 5, unbroken partition between two capsules; 6, cavity resulting from the fusion of several cells; it contains young cells (cells of

fœtal marrow) and a great deal of free fat; 7, angle corresponding to position of a partition which has disappeared.

FIG. V. **Ossification by periosteum** (femur of a newly
born infant). 1, Completely formed bone; 2, deepest portion
of the periosteum, in which some connective fibres and a large
number of plasmatic cells can be still distinguished; of the latter, those nearest the bone begin to resemble bone corpuscles
in shape; 3, superficial portion of periosteum, show few plasmatic cells and numerous connective fibres; 4, plasmatic cells.

PLATE VII.

BONE, *continued;* TEETH.

FIG. I. **Ossification of the os frontis** at the margin of the
anterior fontanelle (from an infant four months old); 1, recently
formed bone; 2, deepest portion of the periosteum; 3, superficial portion of the periosteum. The plasmatic cells, in this
specimen, give off distinctly marked branches.

FIG. II. **Ossification of cartilage**, according to the most
generally received theory.—1, Capsule and cells, unaltered;
first appearance of the corrugation of the cell-wall; 3, corrugation more marked; 4 and 5, corrugation still farther advanced
and completed, resulting in formation of a bone-corpuscle.

FIG. III. **Transverse section of the canaliculi of the
ivory of a tooth.**—1, Canaliculi; 2, their anastomotic
branches; 3, canaliculi divided a little obliquely.

PLATE VIII.

TEETH, *continued.*

FIG. I. **Incisor tooth** of a child nine years old—magnified
thirteen diameters.—1, Dental cavity; 2, ivory; 3, cementum
investing its root; 4, enamel covering its crown.

Fig. II. **Ivory and cementum.**—1, Amorphous substance; canaliculi of the ivory, with their lateral anastomotic branches; 3, dilatations in the course of the canaliculi; 4, confluence of several canaliculi; 5, inter-globular spaces; 6, cementum with very large bone-corpuscles. Some of these latter communicate with the cavities of the inter-globular spaces.

Fig. III. **Transverse section of the crown** of a large molar tooth.—1, Ivory, and terminations of its canaliculi; some of these canaliculi enlarge in diameter (2) and penetrate the substance of the enamel; 3, enamel, consisting of wavy and parallel prisms; they are seen in groups slightly diverging from each other; 4, lines of separation between the prisms.

Fig. IV. **Transverse section of enamel.**—1, Prisms seen in transverse section; prisms divided a little obliquely. The white lines are the intervals between the prisms.

———

PLATE IX.

MUSCLE.

Fig. I. **Muscular coat of the stomach** treated by acetic acid.—1, Finely granulated muscular fibre, with very pale outlines, often indistinctly visible; 2, nuclei; 3, lines of separation between the muscular fibres; 4, elastic fibres.

Fig. II. **Same coat** in transverse section, and hardened by moderate boiling.—1, Muscular fibres; 2, nuclei; 3, line of separation between the fibres; 4, outline of a fasciculus of muscular fibres.

Fig. III. **Dartos** treated by acetic acid.—1, Very pale finely granulated substance, corresponding to the muscular fibres; 2, elongated nuclei.

Fig. IV. **Embryonic fibres** of striped muscle.—1, Two varicose fibres formed by the union of embryonic cells; 2, nuclei of these cells; 3, two other fibres, a little longer and less varicose; 4, division of the contents of the cells into granules and transverse striæ; 5, fibres showing the commencement of striation in the direction of their length; 6, another fibre, in which this appearance is more strongly marked.

FIG. V. **Gemellus muscle** hardened by cooking (from a newly born infant).—1, Myolemma; 2, its contents, showing transverse striæ; 3, nucleus; 4, a broken fibre, with its contents divided into discs; 5, a fibre in which the division of its contents into discs is very well marked.

PLATE X.

MUSCLE, *continued*.

FIG. I. **Antero-posterior section of the tongue** (in a newly-born infant).—1, Muscular fasciculi seen in the direction of their length; 2, same, in transverse section.

FIG. II. **Different views of striped muscular fibre.**— 1, A fibre, the contents of which are crushed in two places; at its left extremity its myolemma is very well seen, corrugated and contracted upon itself; 2, a fibre with transverse striæ, showing a nucleus (3); 4, a fibre in which both longitudinal and transverse stripes are visible; 5, another fibre broken off at its upper extremity; each fibrilla is seen to be composed of a series of slightly flattened granules superimposed upon each other. All of these muscular fibres were procured from the perfectly fresh biceps muscle of a suicide.

FIG. III. **Fibres of the heart.**—1, A common trunk giving off several branches; 2, divisions of the trunk.

PLATE XI.

Distribution of the nerves as seen in the subcutaneous pectoral muscle of a frog. The parallel lines indicate the outlines of the muscular fibres.

PLATE XII.

ELEMENTS OF NERVE TISSUE.

FIG. I. **Nerve fibres.**—1, Nerve fibres of the large variety ; 2, envelope of the fibres; 3, its contents; 4, another fibre treated by chromic acid; 5, envelope; 6, medulla; 7, axis cylinder; 8, fine nerve fibres with a single outline, taken from the spinal marrow.

FIG. II. **Fibres of Remak** taken from a sympathetic ganglion from the lumbar region.

FIG. III. and IV. **Connexion between nerve-fibres** and ganglionic cells (after Leydig).

FIG. V. **Connexion between nerve-fibres and cells** of the spinal marrow. 1, Central canal of the medulla spinalis ; 2, nerve-cells ; 3, superior prolongation ; 4, inferior prolongation ; 5, anterior root ; 6, posterior root ; 7, transverse prolongation forming the anterior commissure and establishing anastomoses between the cells of the two halves of the spinal marrow (after Owsjanikow).

FIG. VI. **Nerve-cells.**—1, Cells with one prolongation (unipolar) from a dorsal ganglion of the sympathetic; 2, another cell from the ganglion of Gasser (on the fifth cranial nerve) ; 3, mass of pigment.

PLATE XIII.

ELEMENTS OF NERVE TISSUE.

FIG. I. **Nerve cells.**—1, Simple (apolar) cells from the grey matter of the brain ; 2, multipolar cells from the grey matter of the cerebellum ; 3, cell from the floor of the fourth ventricle ; 4, another cell from the grey matter of the cord in the cervical region ; 5, nucleus ; 6, mass of pigment surrounding the nucleus ; 7, two cells from the ganglion of Gasser ; one of them has a nucleated envelope (8).

' FIG. II. **Grey matter from the cerebellum.**—1, Apolar
cells; 2, mass of nuclei grouped around the cells; 3, fine nerve
fibres—varicose.

FIG. III. **Superior cervical ganglion.**—1, Nerve cells
imbedded in a faintly fibrillated substance containing nuclei
similar to those represented on the cell at (8).

PLATE XIV.

TERMINATION OF NERVE-FIBRES.—ARTERIES.

FIG. I. **Nerve-fibre** from the subcutaneous pectoral muscle of the frog; 1, muscular fibre; 2, nerve fibre; 3, terminal
ramifications.

FIG. II. **Pacinian corpuscle.**—1, Its pedicle; 2, its cortical substance divided into lamellæ by concentric lines, on the
concave surface of which numerous little nuclei are seen projecting; 3, its central cavity filled with finely granular matter,
and a considerable number of nuclei with pale outlines; 4, the
nerve fibre which forms the axis of its pedicle, running into its
central cavity, where it ends in a slight enlargement.

FIG. III. **Termination of a nerve fibre of the retina**
(after H. Muller).—1, Nerve-cells; 2, fibres from the optic
nerve; 3, another fibre on its way to rejoin the club-shaped
bodies of the external layer of the retina.

FIG. IV. **Transverse section of the primitive carotid
artery** of a child 15 years of age—magnified 120 diameters.—
1, Internal coat; 2, middle coat; 3, external coat.

FIG. V. **The same section** treated by acetic acid and examined with a magnifying power of 400 diameters.—1, Internal
coat; the transverse section of the elastic fibres of which it is
composed are visible; 2, middle coat; 3, nuclei of muscular
fibres; there is a pale line (4) on each side of the nuclei, indicating the limits of the muscular fibres; 5, elastic fibres; 6,
same, in transverse section.

FIG. VI. **The same artery.**—1, Middle coat; 2, external
coat, composed of elastic fibres, most of which run longitudi-

nally, and which are more numerous and closer together towards its internal limit. Between these elastic fibres are connective fibres which grow pale and break down under the action of acetic acid, leaving a hyaline mass (3).

Fig. VII. **Middle coat** of a branch of the artery occupying the fissure of Sylvius; it is composed entirely of muscular fibres, no traces of elastic fibres being visible.

Fig. VIII. **Four muscular fibres** from the basilar artery. The two on the right have been subjected to the action of acetic acid, by which they are rendered pale, and their nuclei much more distinct.

PLATE XV.

ARTERIES, *continued.*

Fig. I. **Epithelial layer of the internal coat** (from the radial artery).—1, Nucleus; 2, internuclear substance composed of cells the outlines of which are not visible.

Fig. II. **Epithelial cells,** isolated (from the radial artery).

Fig. III. **Fenestrated layer.**—1, Amorphous material through which the fibres of the subjacent coat are visible; 2, elastic fibres imbedded in the amorphous substance; 3, openings or fenestra, of various shapes and sizes; 4, irregular line showing where the layer has torn or ruptured; 5, the subjacent layer consisting of longitudinal elastic fibres (from the radial artery).

Fig. IV. **Longitudinal section of the primitive carotid artery** of a young subject (15 years of age); 1-2, internal coat; 2-3, middle coat; 4, muscular fibres of the middle coat; 5, their nuclei; 6, network of elastic fibres; 7, transverse section of these same fibres.

Fig. V. **Same artery,** dried like the preceding, and treated by acetic acid.—1, Line of junction of its middle and external coats. The latter (2) consists of a web of elastic fibres, which run mostly parallel with the axis of the vessel, and are crossed by connective fibres; these have been rendered invisible by the acetic acid.

Fig. VI. **A recent specimen of the external coat of an artery** simply spread out upon the glass.—1, Elastic fibres; 2, fasciculi of connective fibres.

Fig. VII. **A small artery of the brain,** measuring $\frac{1}{40}$th of a line in diameter, treated by very dilute acetic acid.—1, External coat consisting of connective fibres; 2, transverse muscular cells; 3, their nuclei; they form the middle coat. Beneath this, oval nuclei can be distinguished (4) with their long diameters in the direction of the axis of the vessel. They are imbedded in a thin layer of amorphous substance, and constitute the internal coat.

Fig. VIII. **Two capillaries from the brain,** the upper measuring $\frac{1}{250}$th and the lower $\frac{1}{250}$th of a line in diameter.—1, Structureless wall; 2, nuclei contained in the thickness of this wall; 3, cavity of the vessel.

———

PLATE XVI.

VEINS.

Fig. I. **Capillary,** measuring $\frac{1}{400}$ of a line in diameter; 2, minute vein, measuring $\frac{1}{250}$th; their walls are formed by connective fibres running lengthwise of the vessel and studded with numerous plasmatic cells (3).

Fig. II. **Transverse section of the femoral vein**—first dried and then treated by acetic acid.—1–2, Internal coat; 2–3, middle coat; 3–4, external coat. In the internal coat some of the elastic fibres, of which it is constituted, are seen in longitudinal, and others in transverse section; 5, strata of muscular fibres very well characterized by their club-shaped nuclei (6), the outlines of which are clear and distinct; 7, other muscular fibres, seen in transverse section, most of them having a nucleus (8); 9, strata of elastic and connective fibres alternating with the strata of muscular fibres. The external coat is similar to that of the arteries.

Fig. III. **Valve from the internal Saphœna vein.**—

1, Epithelium; the oval nuclei of its cells only are visible, the outlines of the cells themselves being too pale; 2, subjacent layer, formed exclusively of regularly undulating connective fibres.

Fig. IV. **Elastic sub-epithelial membrane** from a small mesenteric vein, treated by acetic acid.—1, Web of elastic fibres; 2, openings of different dimensions, giving this layer the same appearance as the fenestrated coat of an artery; 3, nuclei of its muscular coat, seen by transmitted light.

Fig. V. **A mesenteric vein** of $\frac{1}{16}$th of a line in diameter.—1, Its external coat, made up of elastic fibres, connective fibres, and plasmatic cells (2); 3, middle coat, entirely muscular; 4, the cells which form its muscular fibres in transverse section, containing nuclei; 5, nuclei of same cells, seen in the direction of their length; 6, elastic membrane of the internal coat, rendered visible by the transparency of the specimen.

PLATE XVII.

VEINS.—LYMPHATIC VESSELS.—GLANDS COMPOSED OF CLUSTERED
FOLLICLES.

Fig. I. **Longitudinal section of the femoral vein,** dried and then treated by dilute acetic acid.—1, Internal coat—its elastic fibres almost all running parallel with the axis of the vessel; 2, middle coat; 3, longitudinal elastic fibres; 4, transverse elastic fibres; 5, muscular fibres irregularly distributed; 6, their nuclei; 7, external coat, made up of mingled elastic and connective fibres; these latter, in consequence of the action of acetic acid, present the appearance of a homogeneous granular mass (8).

Fig. II. **Transverse section of a lymphatic vessel** of the thigh, treated by dilute acetic acid. Its internal coat seems to be composed of but a single layer of epithelial cells.—1, Middle coat, formed entirely of muscular fibres, the nuclei of which are very well seen (2); 3, elastic fibres—of which there are very few; 4, external coat, made up of connective, elastic, and muscular fibres; the latter (5) run parallel with the axis of the vessel.

FIG. III. **A longitudinal section** of the same lymphatic.— 1, Middle coat; 2, muscular fibres seen in transverse section; 3, nuclei; 4, external coat; 5, nuclei of muscular fibres encircling the vessel.

FIG. IV. **Recent specimen** of a valve treated by acetic acid.—1, Nuclei of epithelial cells; 2, elastic fibres; 3, nuclei of muscular fibres.

FIG. V. **Section of a lobe of the sub-lingual gland**— magnified 80 diameters.—1, Excretory duct; 2, its radicles, one of which belongs to each lobule; 3, cavity of one of the cœcal pouches of which the gland is composed; 4, connective tissue surrounding its walls.

PLATE XVIII.

GLANDS COMPOSED OF CLUSTERED FOLLICLES.

FIG. I. **Three cœcal pouches of the sub-lingual gland,** lined by thin epithelium. The nucleus (1) almost fills each of the epithelial cells.

FIG. II. **Sebaceous follicle** from the *scrotum*.—1, Cavity of the follicle filled with cells; 2, young cells containing nuclei, and immediately in contact with the walls of the cavity; 3, other cells, of greater age, in process of fatty transformation; 4, excretory duct filled with minute globules of fat; 5, connective fibres enveloping the follicle; 6, epidermis.

FIG. III. **Sebaceous gland** from the external auditory canal.—1, Body of the gland presenting irregular pouched projections (2); 3, excretory duct.

FIG. IV. **Cells from a sebaceous gland** showing different stages of fatty infiltration.

FIG. V. **Epithelium from a Meibomian gland**—1, Young cells; 2, older cells, filled with oil globules.

Fig. VI. **Milk from the human female.**—1, Milk of the first day; 2, *colostrum* corpuscles; 3, free oil globules; 4, milk on the sixteenth day after delivery—from the same woman.

PLATE XIX.

CLUSTERED GLANDS, *continued.*—TUBULAR GLANDS.

Fig. 1. **Portion of dried and inflated lung,** seen with a magnifying power of 25 diameters.

Fig. II. **Pulmonary vesicles** from a fresh lung.—1, Walls of the vesicles; 2, layer of epithelium lining the walls of the vesicles.

Fig. III. **Pulmonary epithelium** from a fœtus at the third month.—1, Cells in their normal relation; 2, detached cells.

Fig. IV. **A sweat gland,** seen under a magnifying power of 165 diameters (from the palmar surface of the middle finger). 1, Excretory duct lined by its epithelium; 2, nuclei of the epithelium; 3, commencement of the excretory duct; 4, fibrous stroma (connective) of the gland, showing numerous plasmatic cells (5).

Fig. V. **Excretory duct** of the same gland.—1, Its external layer consisting of connecting tissue; 2, its internal layer—structureless basement membrane; 3, polygonal epithelium.

Fig. VI. **Same duct seen in transverse section.**—1, Wall of the duct; 2, its epithelium; 3, its cavity.

Fig. VII. **Transverse section of the excretory duct of a ceruminous gland.**—1, Its walls, showing plasmatic cells; 2, its contents; 3, young cells, such as line the walls of the gland; 4, cells a little more advanced in age.

PLATE XX.

TUBULAR GLANDS, *continued.*—KIDNEYS.

FIG. I. **Portion of the kidney of a cat**—magnified 50 diameters.—1, Straight tubes of the medullary substance; 2, tortuous tubes of the cortical substance; 3, Malpighian tufts.

FIG. II. 1 and 2, Fresh tubes showing their internal epithelial lining; 3, a tube, throughout the greater portion of which (4) its external wall only is visible—contracted and slightly wrinkled; 5, detached epithelial cells; 6, transverse section of an urinary tubule; 7, its epithelium; 8, its cavity.

FIG. III. **Portion of an injected kidney** (from Dr. Bœckel), magnified 60 diameters.—1, Arteries; 2, Malpighian tufts; 3, afferent vessel; 4, efferent vessel; 5, vascular plexus of the cortical substance; 6, same, of the medullary substance.

FIG. IV. **Diagrammatic representation of the structure of the kidney.**—1, A straight tube of the medullary substance; 2, tortuous tube of the cortical substance; 3, its termination in a bulbous expansion; 4, an artery; 5, Malpighian tuft; 6, the efferent vessel; 7, capillary plexus; 8, veins in which the vascular plexus pours its blood; 9, relation between the vascular portion of the Malpighian tuft, and the terminal bulbous expansion of a tube; 10, epithelium covering the surface of the Malpighian tuft, and lining the interior of the urinary tube by the terminal bulb of which it is enveloped.

PLATE XXI.

TUBULAR GLANDS, *continued.*—OVARY.

FIG. I. **Section of a testicle** rendered hard by boiling, magnified 50 diameters.—1, External wall of a secreting tubule; 2, its internal tunic; 3, its cavity and epithelium.

FIG. II. **A fresh tubule of the testicle.**—1, Outer coat of its wall; 2, inner coat; 3, polygonal epithelium.

Fig. III. **Epithelial cells from the epididymis.**

Fig. IV. **Human spermatozoa.**—1, Head of a spermatozoon; 2, its caudal prolongation.

Fig. V. **Development of spermatozoa,** as observed in the Guinea pig.—1, Epithelial cell with a solitary nucleus; 2, epithelial cell with two nuclei; 3, the head of the spermatozoon making its appearance in the periphery of a nucleus; 4 and 5, two other cells inclosing a larger number of nuclei in the same stage of development; 6, nuclei in which the caudal prolongation (7) of the spermatozoon is visible; 8, a nucleus with its spermatozoon uncoiled; 9, free spermatozoa.

Fig. VI. **Ovisac.**—1, stroma of the ovisac; 2, *membrana granulosa* of the ovisac; 3, its proligerous disc; 4, *zona pellucida* of the ovule; 5, yelk; 6, germinal vesicle; 7, germinal spot.

Fig. VII. **Ovisac** containing two ovules, 1 and 2.

Fig. VIII. **An ovule in which the process of segmentation has taken place.**—1, *Zona pellucida ;* 2, segmentation of the *vitellus.*

PLATE XXII.

LIVER.—SPLEEN.—THYROID GLAND.

Fig. I. **Vena portæ of the hog**—magnified 50 diameters.—1, A lobule of the liver; 2, interlobular branches of the *vena portæ ;* 3, their subdivisions; 4, capillary network.

Fig. II. **Human vena portæ** (from a child three years of age), magnified 50 diameters.—1, Branches of the *vena portæ ;* 2, their termination in the capillary plexus.

Fig. III. **Intra-lobular vein** of the rabbit—magnified 50 diameters.—1, Boundary of a lobule; 2, trunk of the vein; 3, capillary network.

Fig. IV. **Hepatic cells.**—1, Large cells; 2, small cells.

Fig. V. **Epithelial cells** from the mucous membrane of the gall-bladder.

FIG. VI. **Diagrammatic representation of the minute structure of a lobule of the liver.**—1, Vena portæ; 2, interlobular vein; 3, capillary network (portal plexus); 4, meshes of this network filled with large hepatic cells; 5, biliary duct; 6, prolongations from this duct terminating in blind extremities; 7, epithelium of the biliary duct.

FIG. VII. **Cellular elements of the spleen.**—Splenic cells; 2, epithelial cells from its blood-vessels.

Fig. VIII. **Thyroid body** (adult).—1, One of its cavities; 2, walls of the cavity composed of connective fibres; 3, plasmatic cells; 4, epithelium which has already undergone change.

PLATE XXIII.

THE SKIN.

FIG. I. **Section of the skin** from the palmar aspect of the last phalanx of the index finger—magnified 60 diam.—1, Epidermis; 2, its external or horny layer; 3, internal layer, or *rete mucosum* of Malpighi. Beneath the epidermis the true skin, or derma, is represented, also in two layers; 4, its superficial, 5, its deep layer; 6, papillæ of the derma; 7, a tactile corpuscle; 8, sweat glands; 9, excretory duct of sweat glands; 10, adipose cells.

FIG. II. **Section of the skin** from the palmar surface of the last phalanx of the middle finger.—1, Cells of the horny layer of epidermis destitute of nuclei; 2, polygonal cells of the *rete mucosum:* 3, oval cells, which always form the deepest layer of the epidermis; 4, structureless and transparent boundary line between epidermis and cutis vera; 5, plasmatic cells of derma, mingled with connective and elastic fibres; 6, tactile corpuscle in the interior of a papilla; 7, the nerve fibre with which it is connected; 8, branches of this nerve fibre; 9, plasmatic nuclei enveloped by amorphous material.

FIG. III. **Section of the skin from the scrotum.**—1, Deep epidermic cells loaded with pigment.

Fig. IV. **Transverse section of a papilla of the true skin.**—1, True skin; 2, its limitary border; 3, epidermis.

PLATE XXIV.

NAILS AND HAIR.

Fig. I. **Transverse section of a nail near its root**—magnified 6 diam.—1, The true skin, forming the matrix of the nail; 2, *rete mucosum ;* 3, epidermic structure of the nail; 4, fold of true skin in which the root and sides of the nail are received; 5, surface of this fold continuous with the matrix; 6, *rete mucosum* of neighboring skin, continuous with that of the nail; 7, line of junction of the epidermis of the neighboring integument with the epidermic layer of the nail.

Fig. II. **The same section**—magnified 25 diam.—1, Papillæ of true skin forming matrix of the nail; 2, rete mucosum of nail; 3, its epidermic layer; 4, line of junction of epidermis of neighboring skin with that forming the nail; 5, the only locality at which they are truly continuous : 6, true skin, with its papillæ, of the fold, or groove, lodging the roots and sides of the nail; 7, rete mucosum; 8, papillæ of the derma in transverse section; 9, sweat duct; 10, epidermis.

Fig. III. **Longitudinal section of a nail**—magnified 6 diam.—1, Nail ; 2, derma ; 3, epidermis.

Fig. IV. **A hair from the scrotum** in its follicle, with a sebaceous gland—magnified 50 diam.—1, lower part of the shaft of the hair; 2, its root ; 3, its bulb; 4, epidermis of the hair; 5, its cortical substance ; 6, its medullary canal ; 7, papilla of the bulb ; 8, true skin forming wall of the hair follicle; 9, exterior epidermic layer ; 10, interior epidermic layer; 11, sebaceous gland ; 12, its excretory duct.

Fig. V. **Imbrication of the cells** forming the epidermic layer of the hair.

Fig. VI. **Cells from the same layer** detached and treated by acetic acid.

FIG. VII. **Portion of the shaft of a hair.**—1, Epidermis; 2, cortical substance; 3, medullary canal filled with cells.

PLATE XXV.

HAIRS, *continued.*—MUCOUS MEMBRANE OF THE ALIMENTARY CANAL.

FIG. I. **Cortical substance of a hair** subjected to the action of caustic potash. It is seen to be made up of fusiform bodies, the result, apparently, of metamorphosis of the nuclei of epidermic cells.

FIG. II. **Hair follicle**—magnified 200 diam.—1, External layer; 2, internal layer, and 3, amorphous limitary border of the involuted portion of true skin forming the follicle; 4, external epidermic layer, corresponding to the *corpus mucosum* of Malpighi; 5, internal epidermic layer, corresponding to the external or horny layer of the epidermis; 6, bulb; 7, vascular papilla; 8, medullary substance.

FIG. II. **One of the papillæ circumvallatæ of the tongue**—magnified 25 diam.—1, Section of the principal papilla surmounted by secondary papillæ, 2; 3, epithelium, presenting a smooth surface.

FIG. IV. **A filiform papilla**—magnified 25 diam.—1, Body of the papilla, surmounted by secondary papillæ, 2; 3, epithelium presenting also secondary papillæ, 4.

FIG. V. **A lenticular papilla**—magnified 50 diam.—1, Central orifice leading into a cul-de-sac; around this the capillaries of the mucous membrane are shown.

FIG. VI. **Epithelium of the œsophagus.**—1, Two of its cells, detached.

FIG. VII. **Surface of the mucous membrane of the stomach**—magnified 25 diam.—1, Orifice of a gastric gland.

FIG. VIII. **Compound pyloric gland of an infant**—magnified 250 diam.—1, 1, Its two terminal cul-de-sacs opening into a common outlet.

PLATE XXVI.

MUCOUS MEMBRANE OF THE ALIMENTARY CANAL, *continued.*

Fig. I. **Cul-de-sac of a compound cardiac gland.** Its epithelial cells are larger than those of the simple and pyloric glands, and possess a different shape.

Fig. II. **Mucous membrane of the duodenum**—magnified 25 diam.—1, A conical villus; 2, same, valvular in shape; 3, a compound valve-shaped villus; 4, orifice of Lieberkuhn's glands.

Fig. III. **Mucous membrane of the ileum**—magnified 25 diam.—1, A villus; 2, orifice of Lieberkuhn's glands.

Fig. IV. **A villus covered by its epithelium**—magnified 250 diam.—1, Cells seen with their bases presenting; 2, cells seen obliquely; 3, amorphous coating.

Fig. V. **Epithelium seen on its superficial aspect**—magnified 400 diam.

Fig. VI. **Cells seen in their whole length;** their bases still covered by the amorphous investment.—1, A cell with two nuclei.

Fig. VII. **A villus deprived of its epithelium.**—1, Amorphous or slightly fibrillated substance; 2, a vascular loop; 3, nuclei, most of which seem to belong to capillary vessels.

Fig. VIII. **Villi, injected**—magnified 50 diam.

Fig. IX. **A Brunner's gland**—magnified 50 diam.

Fig. X. **A Lieberkuhn's gland**—magnified 125 diam.—1, Its wall; 2, its epithelium.

Fig. XI. **Orifice of a Lieberkuhn's gland**—magnified 125 diam.—1, Epithelium of the gland forming a radiating crown around the central open space, 2.

Fig. XII. **A solitary gland of the ileum**—magnified 25 diam.—1, Projection of the gland; 2, villi; 3, orifices of Lieberkuhn's glands.

Fig. XIII. **Two solitary ductless glands** injected—magnified 50 diam.—1, Radicles of the meseraic vein; 2, capillaries surmounting the glands.

Fig. XIV. **Mucous membrane of the colon,** showing orifices of Lieberkuhn's glands.

Fig. XV. **Mucous membrane of colon.**—1, Lieberkuhn's glands; 2, orifice situated over the position of a solitary gland.

SUPPLEMENTARY PLATES.

PLATE XXVII.

Fig. I. **Ossification** in a medullary canal. — 1, Newly formed bone; 2, oval nuclei; 3, nuclei with radiating processes; 4, line of junction of the blastema and bone.

Fig. II. **Transverse section of the orbicularis palpebrarum** muscle.

Fig. III. **Meibomian gland.**—1, Common excretory duct; 2, lobules. Magnified 25 diam.

Fig. IV. **Kidney of guinea-pig.**—1, Urinary tubule; 2, its terminal enlargement; 3, Malpighian tuft; 4, afferent and efferent vessels; 5, epithelium covering surface of the tuft. Magnified 240 diam.

Fig. V. **Transverse section of an eye-lash** at its base.— 1, Medullary substance; 2, cortical substance of the hair; 3, internal epidermic layer of the hair follicle; 4, external epidermic layer; 5, internal zone of the derma of the hair follicle; 6, external zone; 7, sebaceous glands. Magnified 220 diam.

Fig. VI. **An internal villus** bare of epithelium, taken from a portion of intestine during the progress of digestion.—1, Body of the villus infiltrated with oil-globules; 2, lacteal occupying the central axis of the villus and ending by a closed extremity. Magnified 220 diam.

PLATE XXVIII.

THE EYE.

FIG. I. **Section of the two external tunics of the eyeball** at the junction of the cornea and sclerotica.—1, Sclerotica ; 2, cornea; 3, line of junction of these two parts; 4, canal of Schlemm ; 6, conjunctiva ; 7, corneal and conjunctival epithelium ; 8, line of junction of conjunctiva and sclerotica ; 9, amorphous layer on the front of the cornea; 10, same layer on posterior surface of cornea ; 11, iris; 12, choroid ; 13, a ciliary process. Magnified 25 diam.

FIG. II. **Cornea, sclerotica, and conjunctiva.**—1, Sclerotica; 2, cornea; 3, amorphous layer on front surface of cornea; 4, junction of this layer with the conjunctiva ; stratified epithelium of the conjunctiva and anterior surface of cornea. Magnified 300 diam.

FIG. III. **Section of cornea and iris.**—1, Cornea ; 2, anastomosing plasmatic cells ; 3, its posterior amorphous layer ; 4, junction of this layer with the sclerotica ; 5, its epithelial investment which, reflected upon the anterior surface of the iris, constitutes the membrane of Demours. Magnified 360 diam.

FIG. IV. **Ciliary muscle.**—1, Sclerotica; 2, canal of Schlemm ; 3, ciliary ring ; 4, ciliary processes in which nuclei of muscular fibres are seen in the direction of their length at 5, and in transverse section at 6 ; 7, external circumference of the iris. Magnified 120 diam.

Pl.1.

Fig.1.

Fig.ll.

Fig.Vl.

Fig.Vll.

Fig.lll.

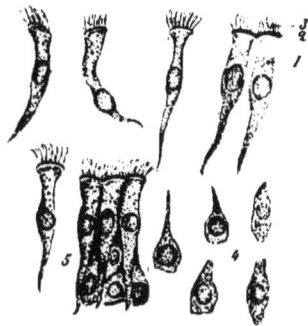

Fig.lV

Fig.V

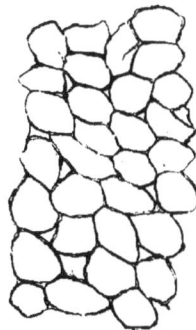

C.Morel prap. Villemin.del.   Lith.E.Simon Strast.9

Pl. II.

Fig. II

Fig. I.

Fig. III

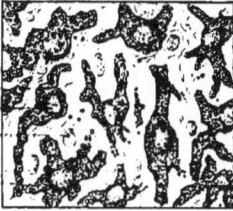

Fig. VI

Fid. IV.

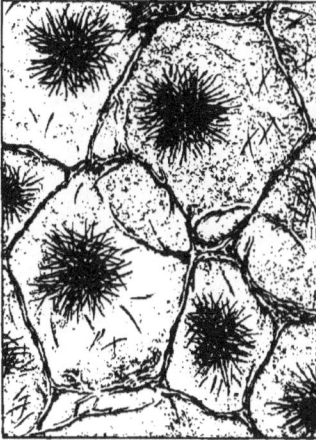

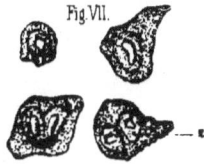

Fig. VII.

Fig. VIII.

Fig. V

Fig. IX

Pl. III

Fig. I.

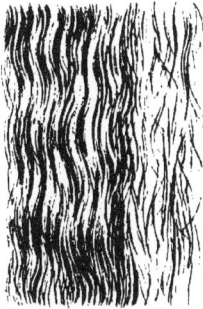

Fig. II.

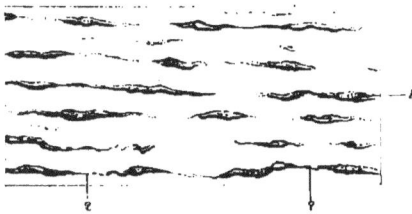

Fig. III

Fig. IV.

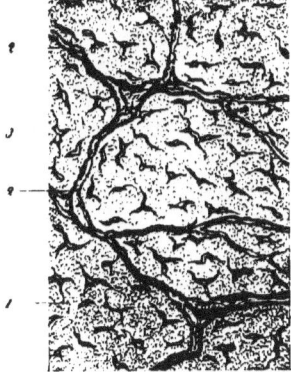

Fig. V

Fig VI.

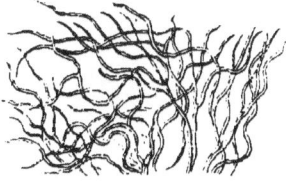

Pl. IV.

Fig. I.

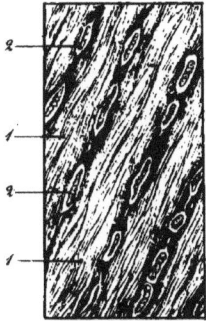

Fig. II.

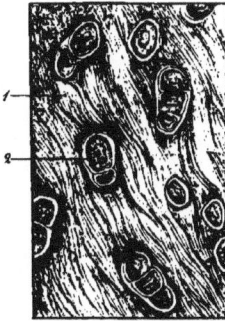

Fig. V.

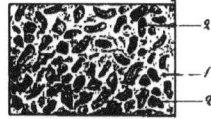

Fig. VI.

Fig. VII.

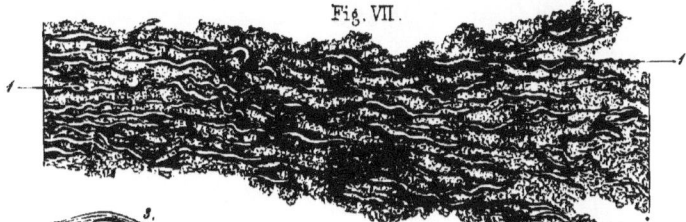

Fig. III.

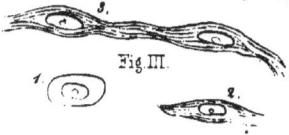

Fig. IV.

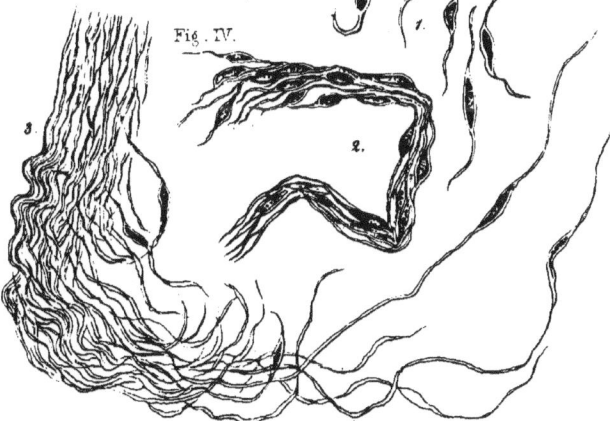

C. Morel prép.-Sillemin del.   Lith. E. Simon.Strasg.

Pl. V.

Fig. 1.

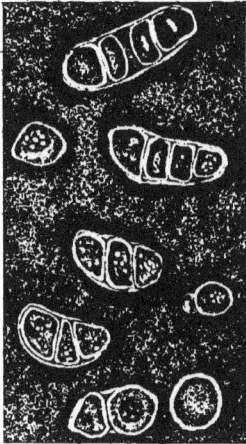

Fig. II.

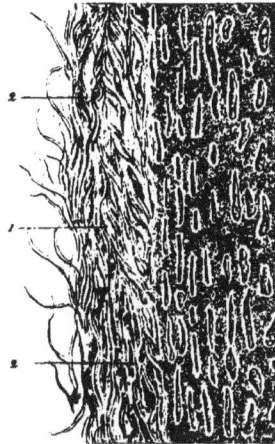

Fig. III.

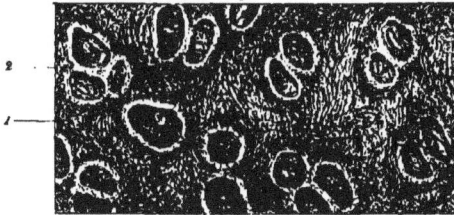

Fig. 2

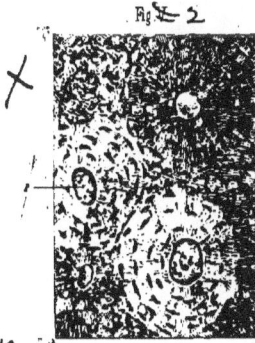

Fig. IV.

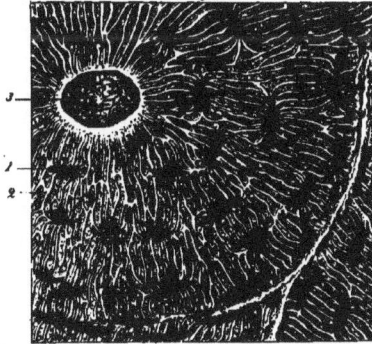

Transverse section of bone
magnified 80 diameters.

C. Morel praep. Villemin. del.   Lith. F. Simon à Strasb.

Pl. VI

Fig. I

Fig. II

Fig. V

Fig. IV

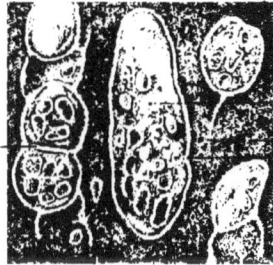

Fig. III

Ollard præp. Villemin del.

Pl VlI

Fig III.

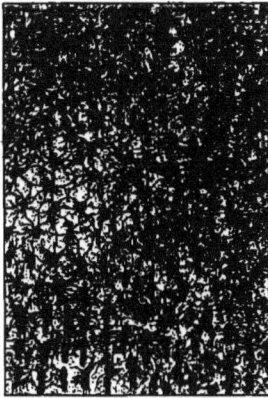

Fig I

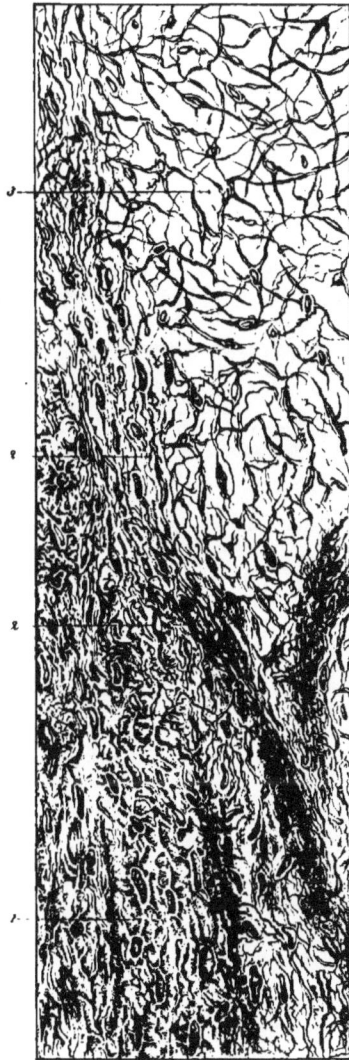

Fig II

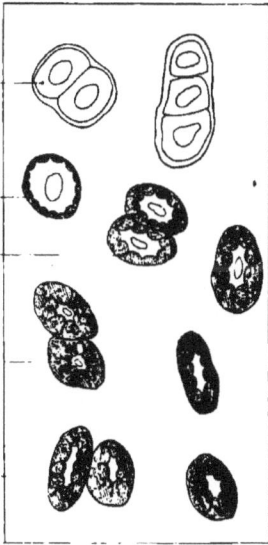

Pl. VIII.

Fig. II.

Fig. III.

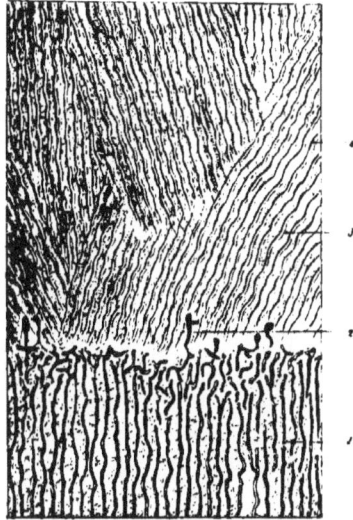

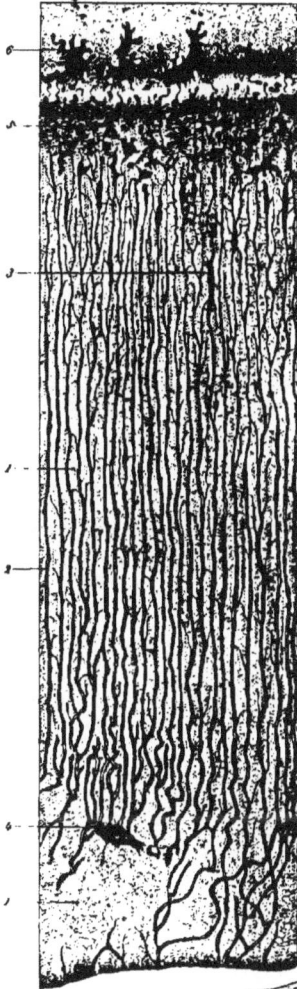

Fig IV.

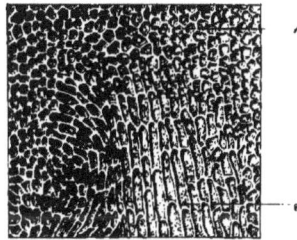

Fig I.

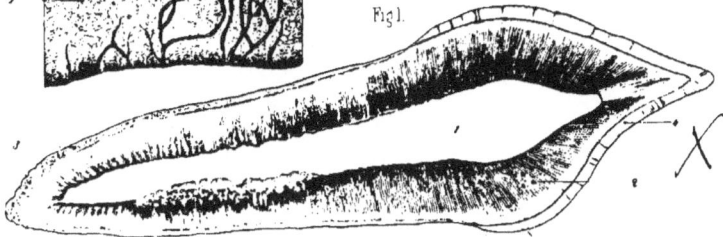

C. Morel prap. Villemin del.

Pl. IX.

Fig. I.

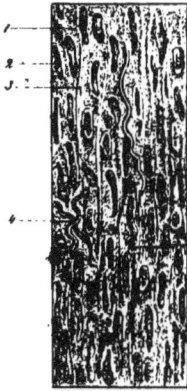

Fig. II.

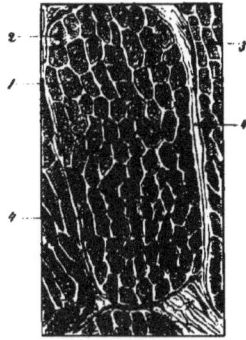

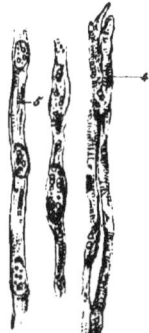

Fig IV.

Fig. V.

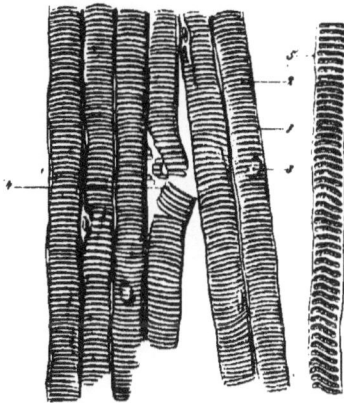

Fig III

C. Morel, prep. Villemin, del.

Pl. X

Fig. 1

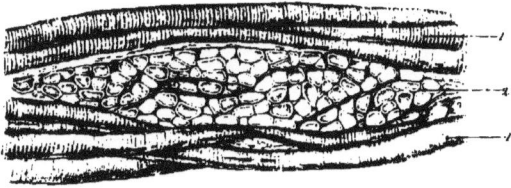

Fig. II.

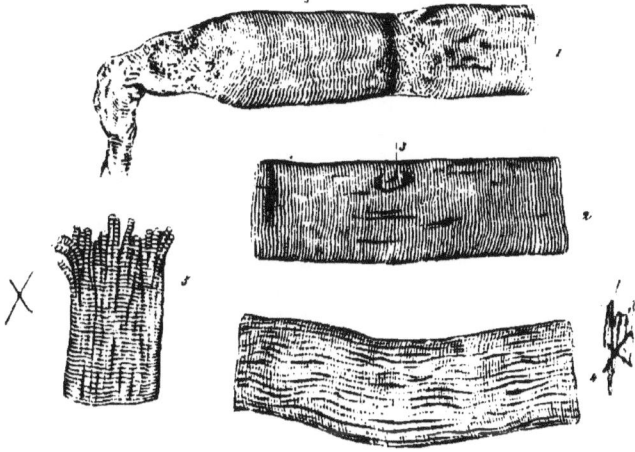

Fig. III.

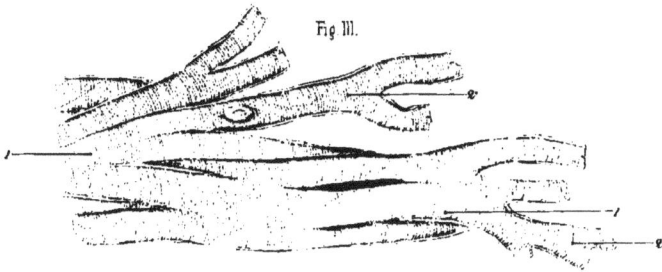

C. Morel prap. Villemin. del. Lith. E. Simon Strasb.

Pl. XI.

C. Morel præp. Villemin del.   Lith.E.Simon à. Strasb.

Pl. XII.

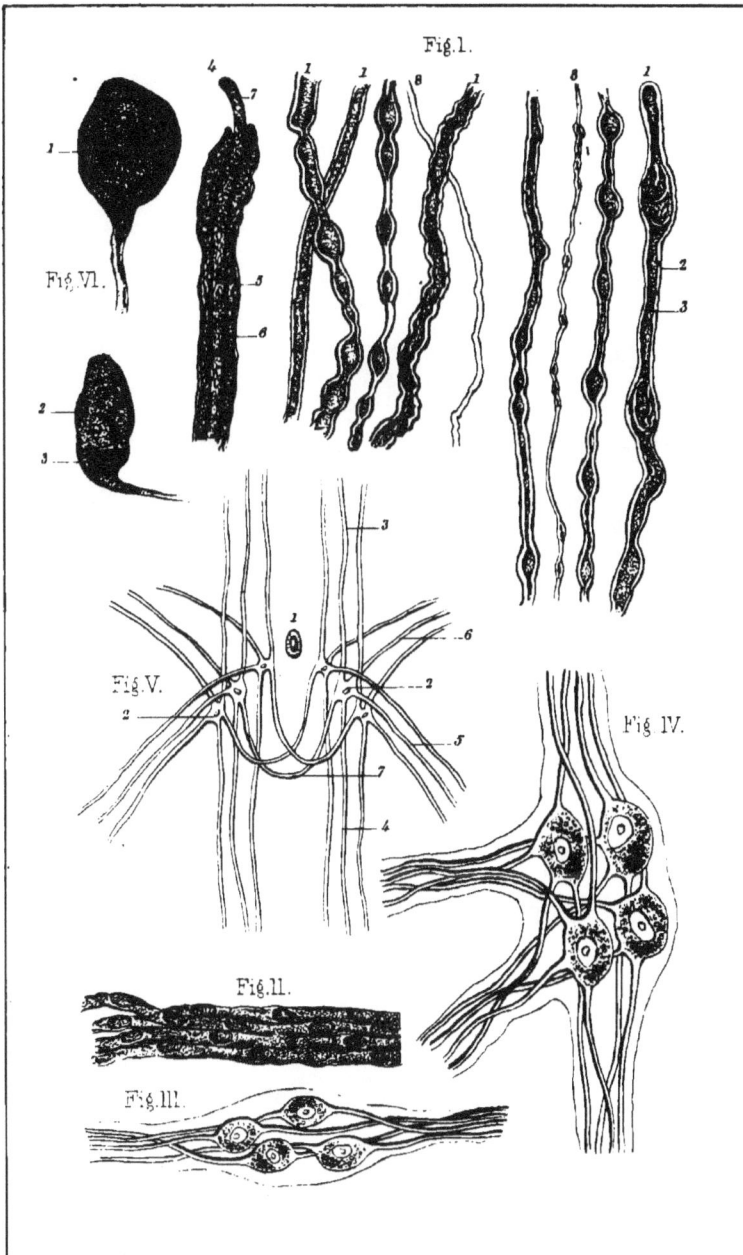

Fig. 1.

Fig. VI.

Fig. V.

Fig. IV.

Fig. II.

Fig. III.

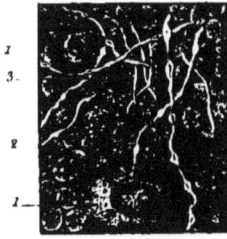

Fig. II.

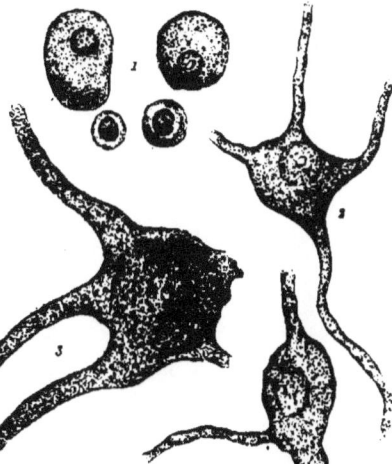

Fig. 1.

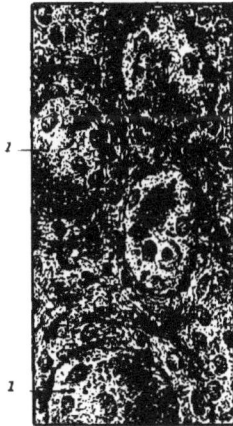

Fig. III.

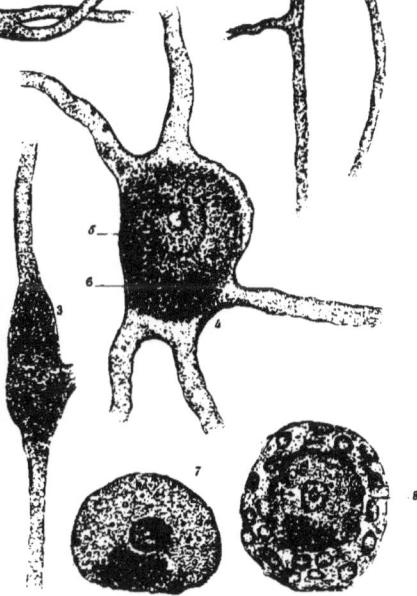

C. Morel prap. Lillemin del.

Pl. XIV.

Fig.I.

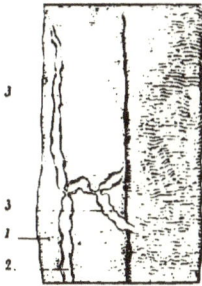

Fig.VI.

Fig.II

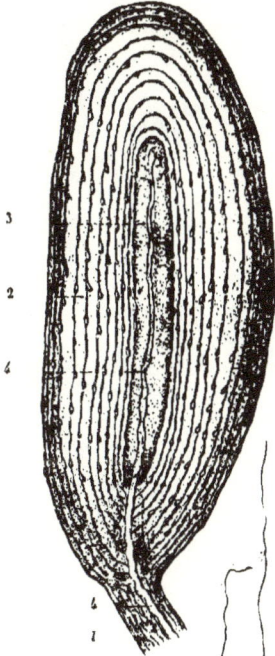

Fig.VII.

Fig.VIII.

Fig.IV.

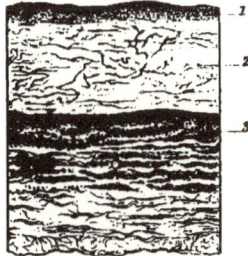

Fig.V.

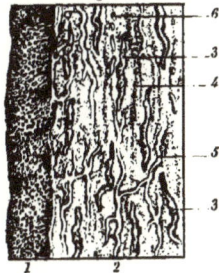

Fig.III

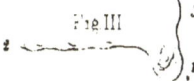

C. Morel præp.Villemin del.   Lith P.Simon Strasbg.

Fig. IV.

Fig V.

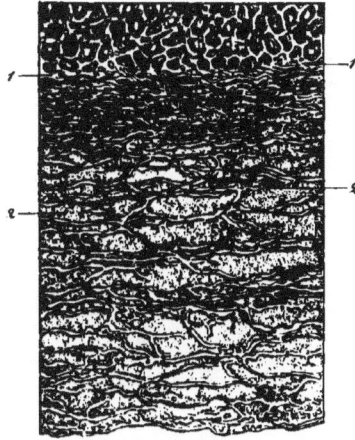

Fig. I.

Fig. II.

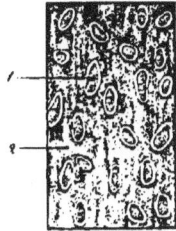

Fig. III.

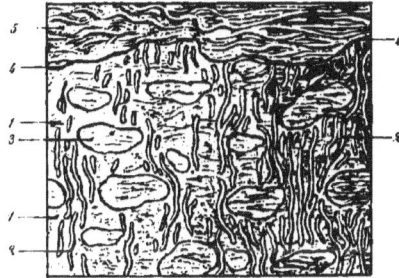

Fig. VI.

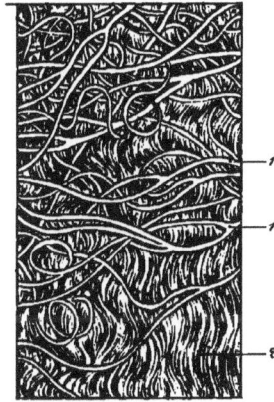

Fig. VIII.

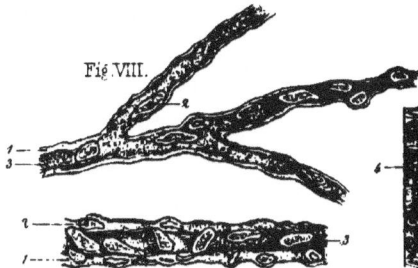

Fig. VII.

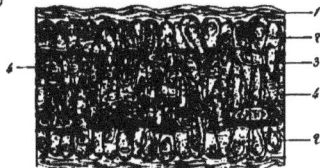

C. Morel præp. Villemin del.

Fig. II.

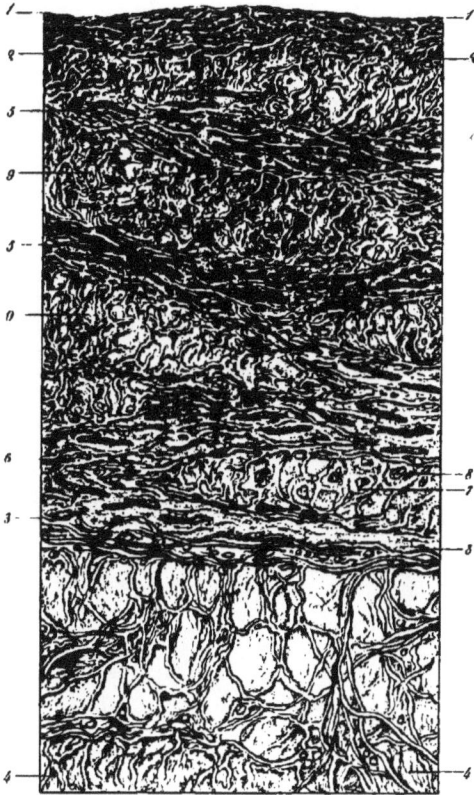

Fig. III.

Fig. I.

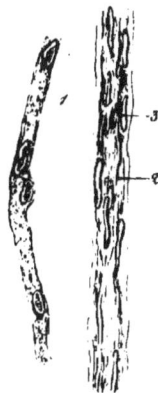

Fig. IV.

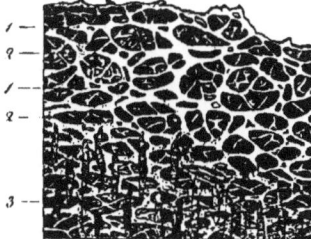

Fig. V.

C. Morel præp. Villemin del.

Pl. XVI.

Fig I.

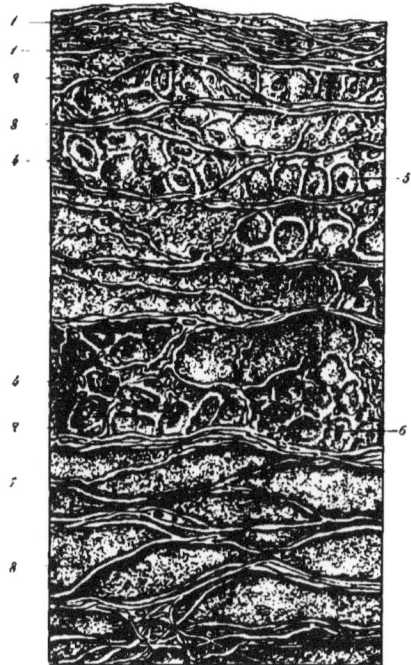

Fig.III.

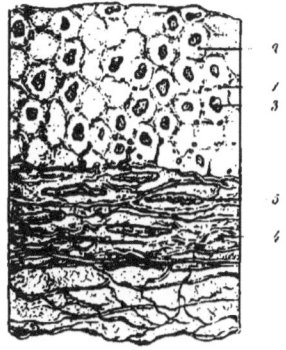

Fig.IV.

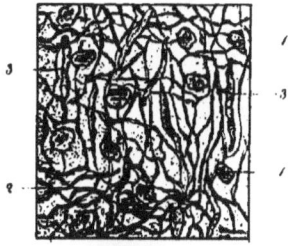

Fig.II.

Fig.V.

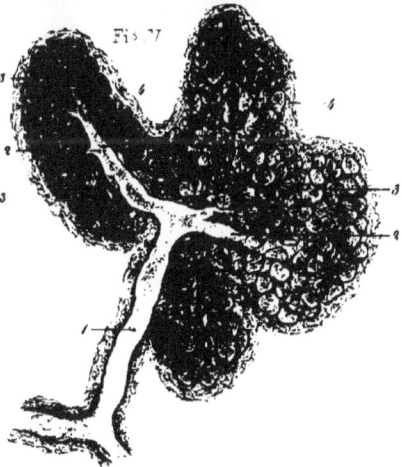

Pl. XVIII.

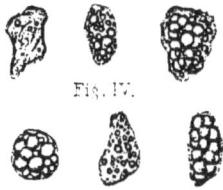

Fig. IV.

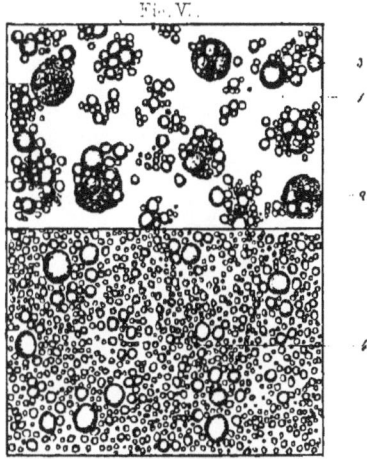

Fig. V.

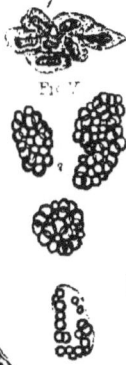

Fig. VI.

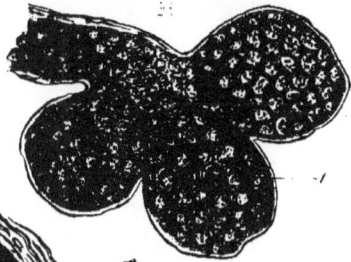

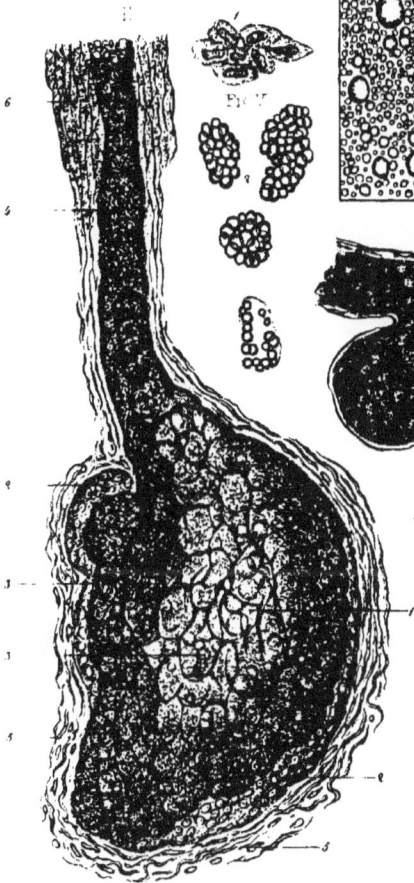

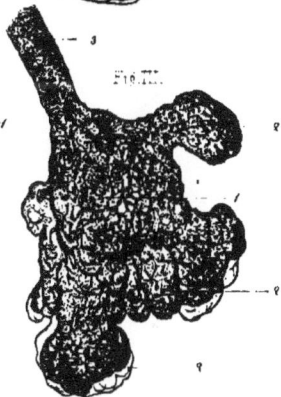

Fig. III.

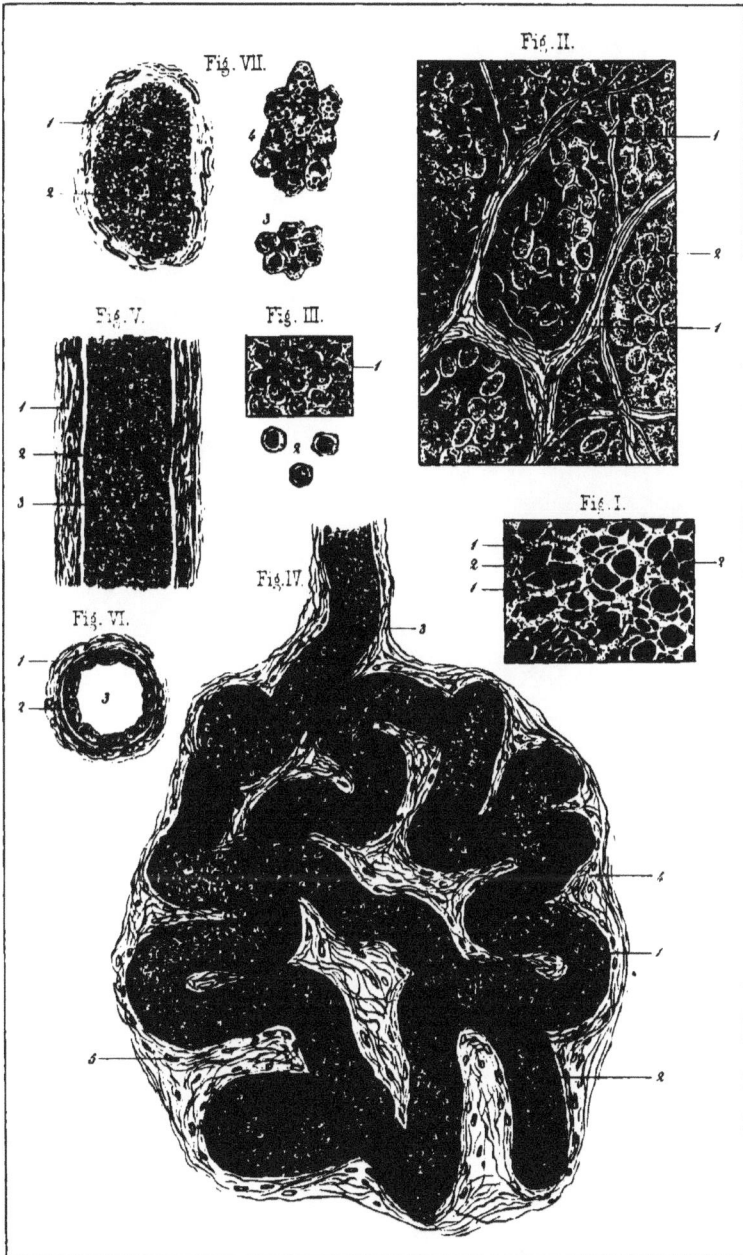

Pl. XIX.

Fig. VII.

Fig. II.

Fig. V.

Fig. III.

Fig. I.

Fig. IV.

Fig. VI.

C. Morel præp–Lillemin del.

Pl. XX.

Fig. I.

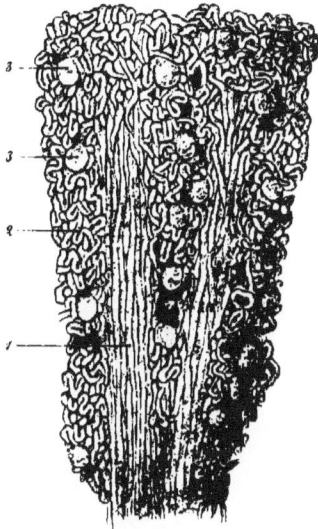

Fig. II.

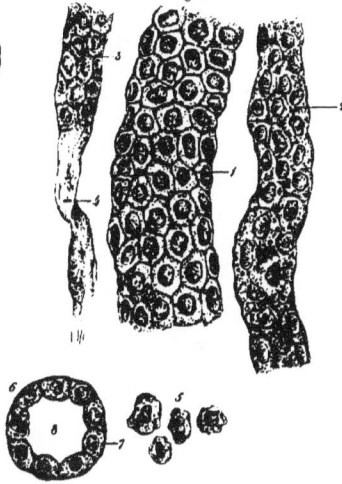

Fig. IV.

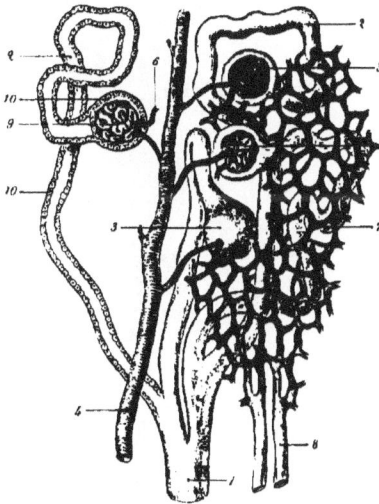

Fig. III.

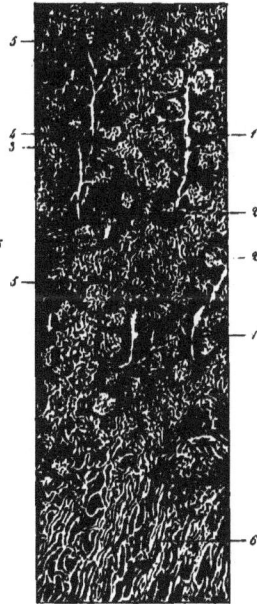

C. Morel præp. Villemin del.

Pl. XXI.

Fig. II.

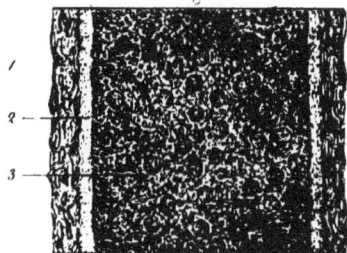

Fig. I

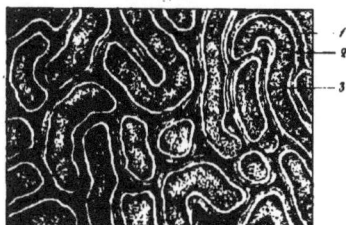

Fig. V.

Fig. IV.

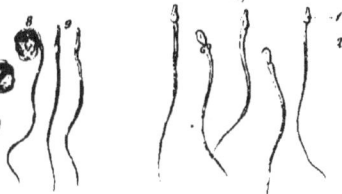

Fig. III.

Fig. VIII.

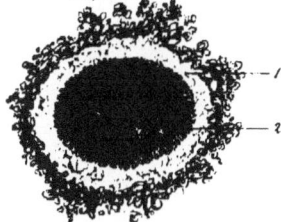

Fig. VI.

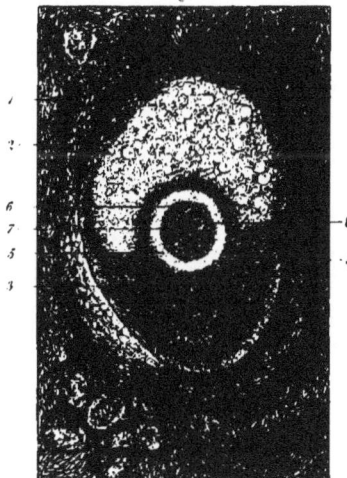

Fig. VII.

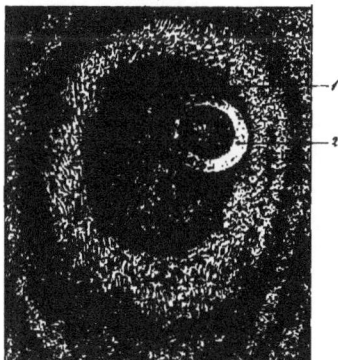

C. Morel prap.–Villemin del.

Pl. XXII.

Fig. III.

Fig. VIII.

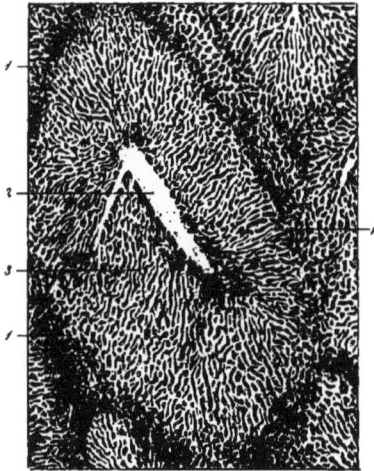

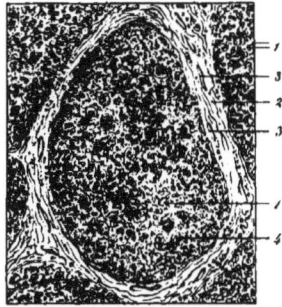

Fig. VII.

Fig. I.

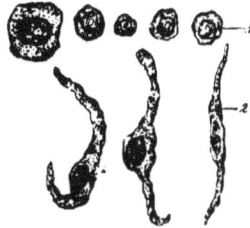

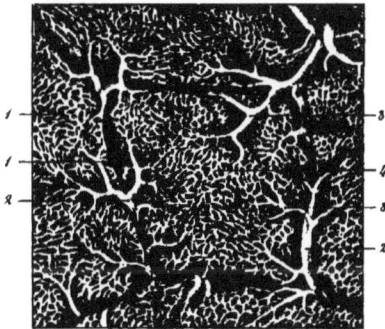

Fig. V.

Fig. VI.

Fig. II.

Fig. IV.

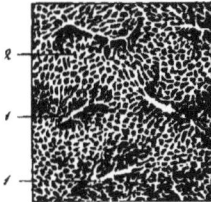

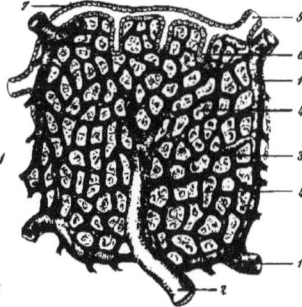

C. Morel præp_Villemin del.   Imp. E. Cremer. &c 1869.

Pl. XXIII.

Fig. I.

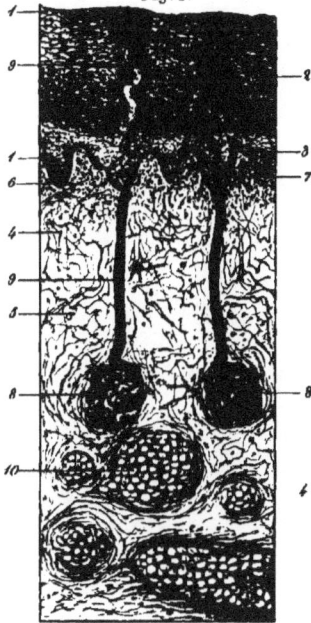

Fig. II.

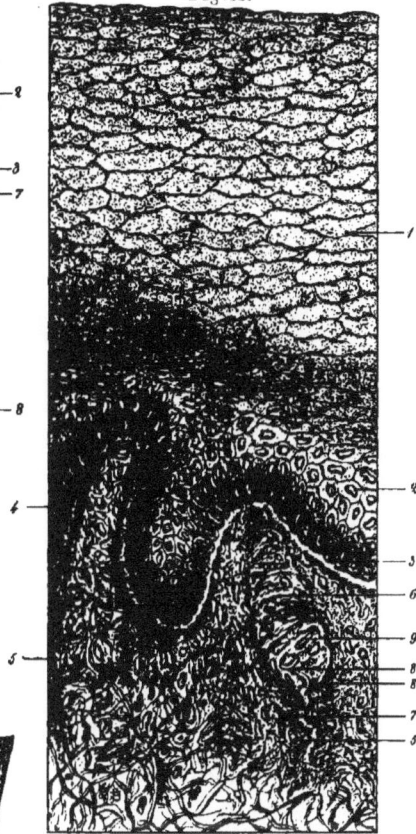

Fig. III.

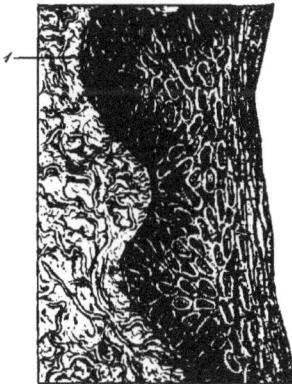

Fig. IV.

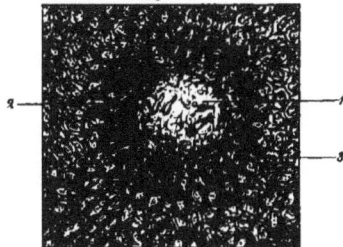

C. Morel præp. Villemin del.   Lith. E. Simon à Strasbg.

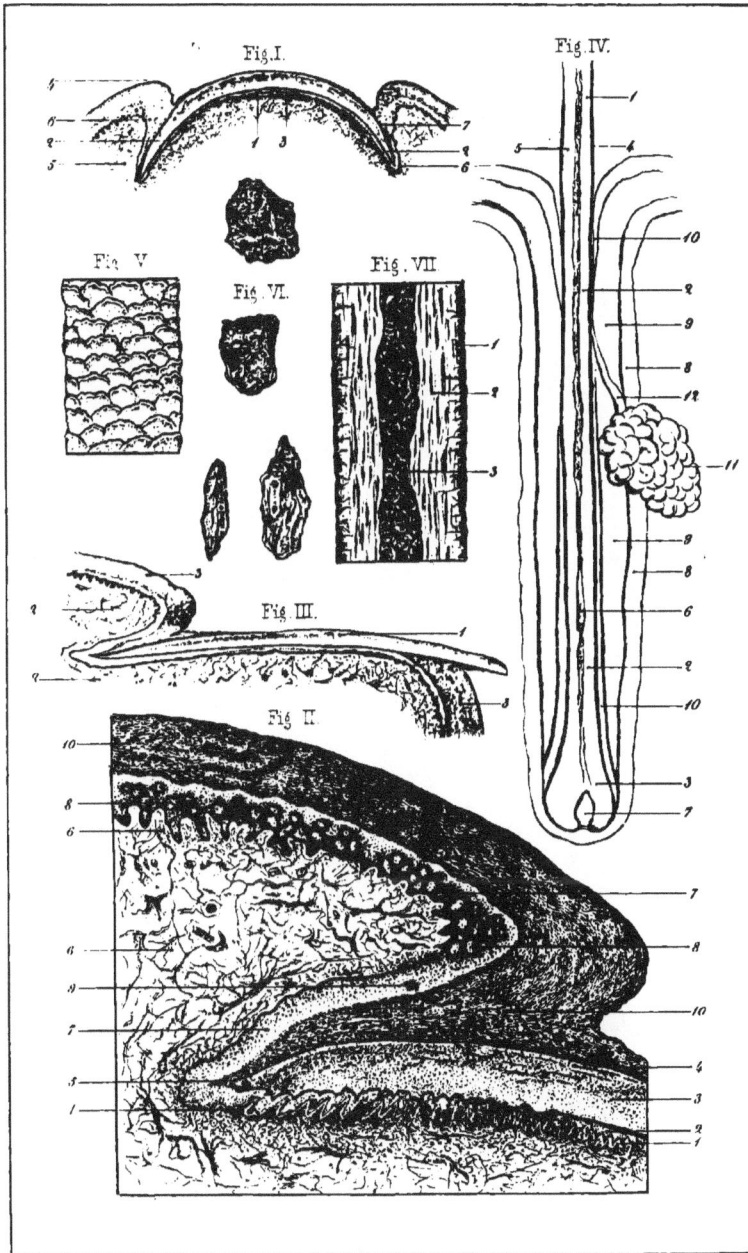

Fig.I.

Fig.IV.

Fig. V.

Fig. VI.

Fig. VII.

Fig. III.

Fig. II.

C. Morel præp.-Lillemin del.

Pl. XXV.

Fig. I.

Fig. II.

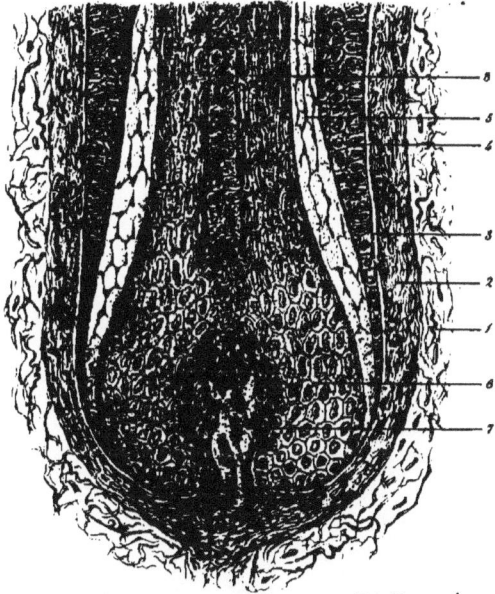

Fig. VIII.

Fig. III.

Fig. V.

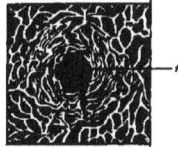

Fig. IV.

Fig. VI.

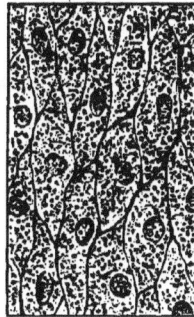

Fig. VII.

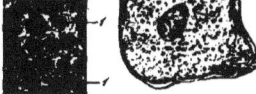

C. Morel præp._ Villemin del.   Imp. F. C. nn. Strasbourg.

Pl. XXVI.

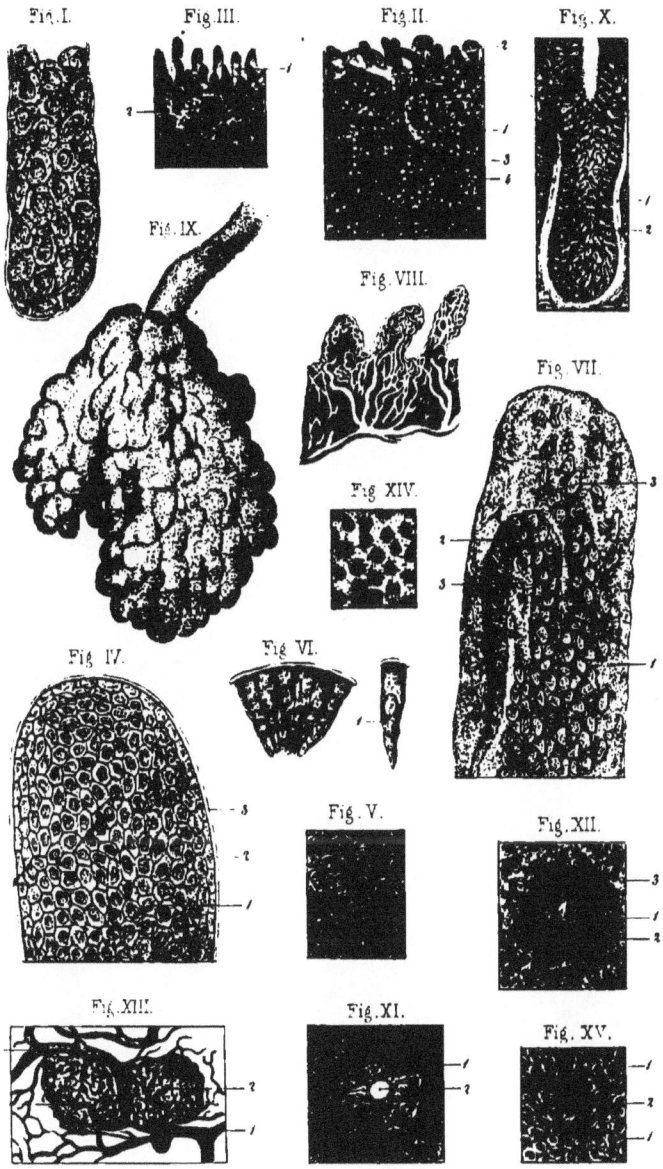

Fig. I.   Fig. III.   Fig. II.   Fig. X.

Fig. IX.   Fig. VIII.   Fig. VII.

Fig. XIV.

Fig. IV.   Fig. VI.   Fig. V.   Fig. XII.

Fig. XIII.   Fig. XI.   Fig. XV.

C. Morel prap. Aillemin del.

Fig.III.

Fig V

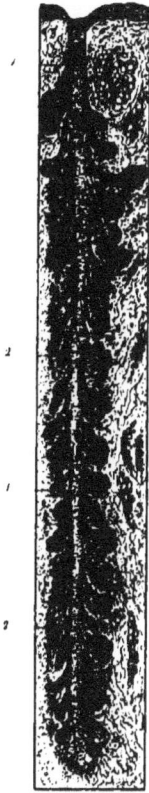

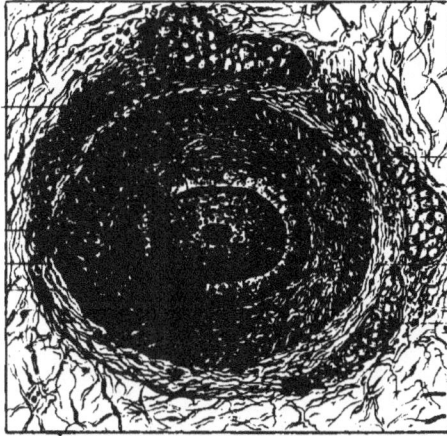

Fig.II.

Fig.I.

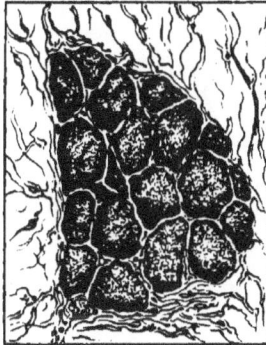

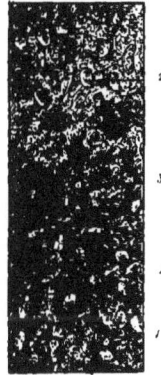

Fig.IV.

Fig.VI.

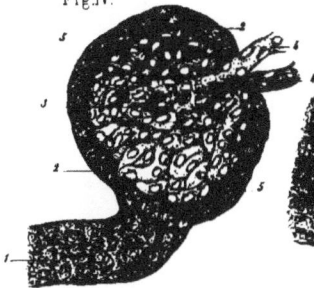

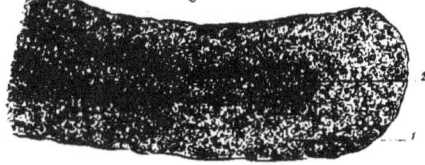

Fig. IV.

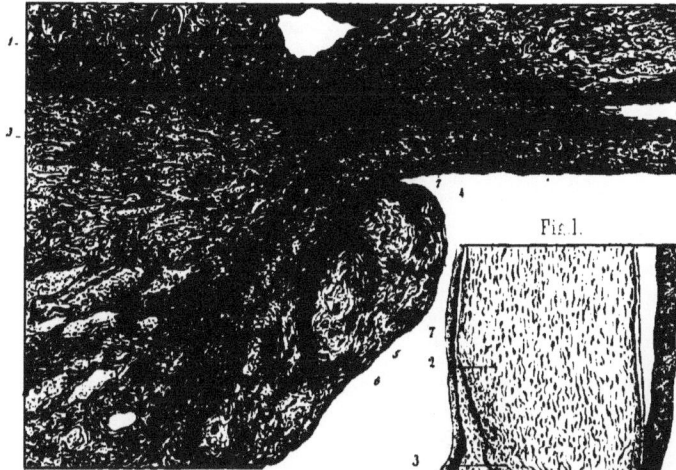

Fig. I.

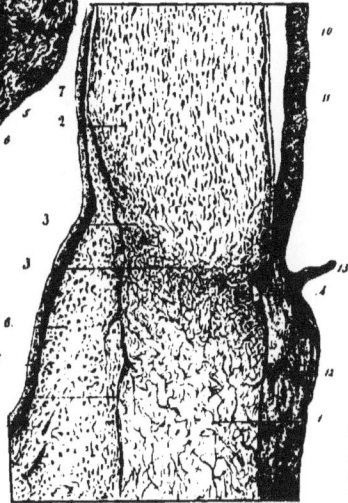

Fig II

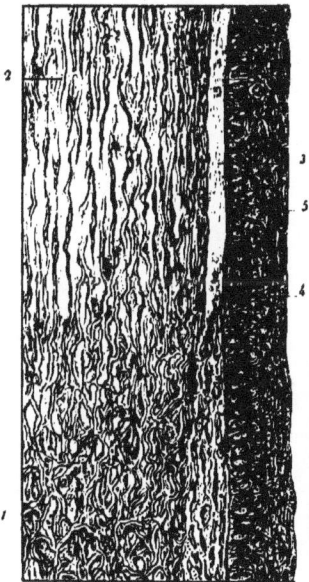

Fig III

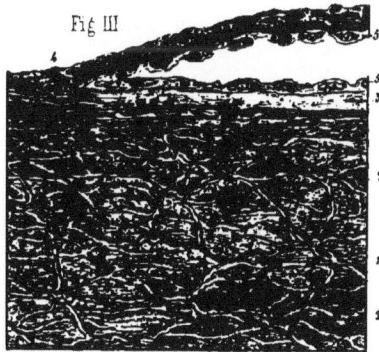

C. Morel præp. Villemin del.

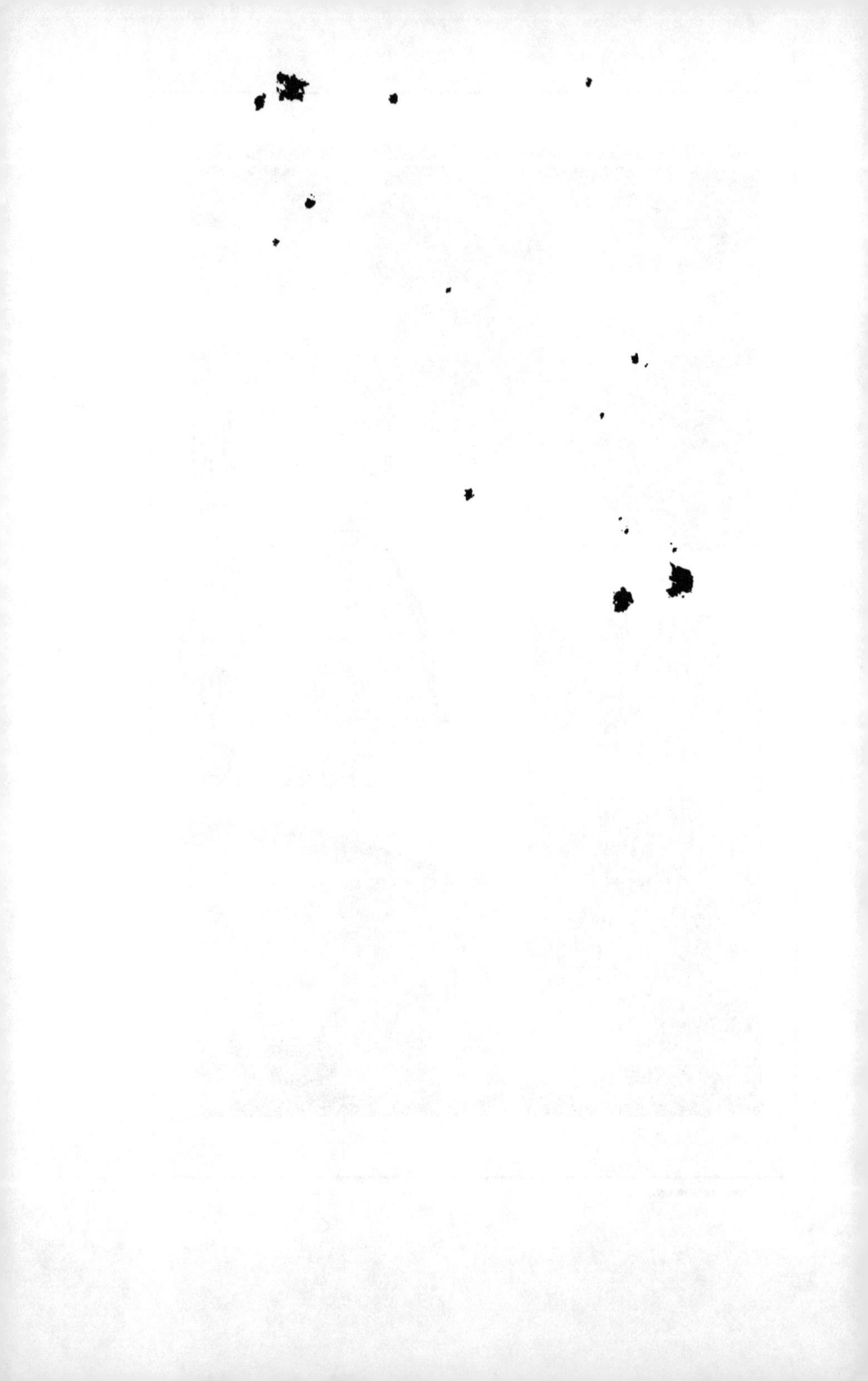

American Medical Times.

A WEEKLY MEDICAL PERIODICAL.

The AMERICAN MEDICAL TIMES is published every Saturday Morning in quarto form, twenty-four pages, double columns, with clear type and on fine paper. It will make two annual volumes, each consisting of over 600 pages. Each number contains Original Lectures and Papers, Reports of Hospitals and Societies, Editorials, Reviews, Reports on the Progress of Medical Sciences, General, Foreign and Domestic Correspondence, Medical News, &c., &c.

** The volumes commence regularly on the first of January and July, but Subscriptions may begin at any date. Subscribers who desire to have the Series complete may, for the present, be supplied with the back Numbers.

TERMS $3.00 a Year in Advance.

BAILLIERE BROTHERS, 440 Broadway, New York.

October 1, 1860.

In establishing the AMERICAN MEDICAL TIMES, the Publishers have sought to meet a manifest and often expressed want, on the part of the medical profession of the city and the country. This want is but the natural result of those changes which have been wrought in society within the last few years, and which are seen in the now universal activity and enthusiasm in the study and investigation of every branch of science, the increased facilities for the immediate and wide diffusion of intelligence, and the rapid growth and expansion of our medical institutions. Scientific medicine now partakes largely of that spirit of inquiry and restless activity which characterizes the age. In every department ardent and impulsive adventurers, with all the aids of modern invention, are enlarging the bounds of knowledge, and daily unfolding truths and principles of the greatest practical utility. The rapid development of every branch of medicine has created a desire on the part of the profession at large, for the earlier and more frequent dissemination of scientific intelligence by the medical press.

It was to meet these requirements that the Publishers proposed to issue a weekly periodical of large size, and adapted to fully represent the progress of medical science, not only in our city and country, but throughout the world. They were at this time proprietors of the New York Journal of Medicine, a bi-monthly which had for seventeen years shared largely the confidence of the profession, and circulated widely throughout the Union. But it could not in that form meet the demands of its readers, nor faithfully represent New York, with her population increased three-fold since its first issue, her numerous and well appointed hospitals, her flourishing schools of medicine, and her gradual centralization of medical interests. Accordingly they combined the interests of that Journal with those of the Medical Press, a weekly medical sheet of limited size, and established the AMERICAN MEDICAL TIMES, which contains in the aggregate THREE TIMES the reading matter of either of these periodicals, though they were of dimensions equal to any now published in this country. That the AMERICAN MEDICAL TIMES may come within the reach of the practitioner of the most limited means, the subscription price is continued at $3.00 per annum.

The general Editorial management of the MEDICAL TIMES is confided to STEPHEN SMITH, M.D., who for many years successfully conducted the New York Journal of Medicine, with whom is associated a corps of experienced collaborators in its several departments.

Specimen Numbers and Detailed Prospectus can be had by addressing

BAILLIERE BROTHERS, 440 Broadway, N. Y.

BEALE (L.) Illustrations of the Salts of Urine, Urinary Deposits, and Calculi. 8vo. London, 1858 . . . . 2 75

———— Use of the Microscope in its application to practical medicine. 2d edition, 8vo. London, 1858 . . . . 4 25

———— How to work with the Microscope, Crown 8vo. London, 1857 . . 1 50

———— Illustrations to How to work with the Microscope. Post 8vo. London, 1859 0 50

———— Tables for the Microscopical Examination of Urine. 8vo. London, 1857 . 0 75

BECQUEREL (L. A.) Traité clinique des Maladies de l'Uterus et de ses Annexes. 2 vols. 8vo. avec Atlas de 18 planches. Color. Paris, 1859 . . . . 5 00

BELL (A. N.) A Knowledge of Living Things, with the Laws of their Existence. 12mo. New York 1 50

BERAUD ET ROBIN. Elements de Physiologie de l'Homme et des Principaux Vertebras. 2eme edition. 2 vols. 12mo. Paris, 1856-57 . . . . 8 00

BERNARD AND HUETTE. Illustrated Manual of Operative Surgery and Surgical Anatomy. Edited with Notes and Additions, by W. H. VAN BUREN, M.D., Professor of Anatomy, University Medical College, and C E. ISAACS, M.D. Complete in one volume, 8vo. with 118 colored plates, half-bound morocco, gilt tops. 1856. . 15 00

———— Plain plates 9 50

BERNARD (CL.) Lecons de Physiologie experimentale appliquee a la Medecine, faites au College de France. 2 vols. 8vo. avec figures. Paris, 1855-56 . . 3 50

———— Cours de Medecine du College de France. Des effets des substances toxiques et medicamenteuses. 8vo. avec figures. Paris, 1857 . . . . 1 75

———— Cours de Medecine du College de France. Lecons sur la Physiologie et la Pathologie du Systeme Nerveux. 2 vols. 8vo. avec figures. Paris, 1853 . . 8 50

———— Lecons sur les Proprietes Physiologiques et les alterations Pathologiques des differents liquides de l'Organisme. 2 vols. 8vo. avec figures. Paris, 1859 . 8 50

———— Memoire sur le Pancreas et sur le role du suc Pancreatique dans les Phenomenes digestifs, particulierement dans la digestion des matieres grasses neutres. 4to. avec 9 planches en partie colorees. Paris, 1856 . . . . . 8 00

DAY (G. E.) Chemistry in its Relations to Physiology and Medicine. 8vo. Illustrated by plates, London, 1860 . . . . 5 00

DAVAINE (C.) Traite des Entozoaires et des Maladies Vermineuses de l'homme et des animaux domestiques. Paris, 8vo. 1860 . . . . . . 8 00

FAU. The Anatomy of the External Forms of Man, for Artists, Painters, and Sculptors. Edited by R. KNOX, M.D., with Additions. 8vo. text, and 28 4to. plates. London, 1849. Plain 6 00

———— Colored plates 10 00

GERBER AND GULLIVER. Elements of the General and Microscopical Anatomy of Man and the Mammalia; chiefly after Original Researches. By PROF. GERBER. To which is added an Appendix, comprising Researches on the Anatomy of the Blood, Chyle, Lymph, Thymous Fluid, Tubercle, with Additions, by C. GULLIVER, F.R.S. 8vo., and an Atlas of 34 plates. 2 vols. 8vo. cloth boards, 1842 . . 6 00

HALL (MARSHALL). On the Diseases and Derangements of the Nervous System, in their Primary Forms, and in their Modifications by Age, Sex, Constitution, Hereditary Predisposition, Excesses, General Disorder, and Organic Disease. By MARSHALL HALL, M.D., F.R.S. L. & E. 8vo. with 8 engraved plates. London, 1841 . . 8 75

———— New Memoir on the Nervous System. True Spinal Marrow, and its Anatomy, Physiology, Pathology, and Therapeutics. 4to. with 5 engraved plates. London, 1843 . 5 00

KOLLIKER (A.) A Manual of Human Microscopic Anatomy. 8vo. 249 Illustrations. London, 1860 . . . . 7 20

LEBERT (H.) Traite d'anatomie pathologique generale et speciale, ou description et iconographie pathologique des alterations morbides, tant liquides que solides, observees dans le corps humain. 2 vols. in folio de texte, et environ 200 planches dessinees d'apres nature, gravees et la plupart coloriees. Paris, 1855-1860.

Le tome 1, texte 760 pages, et tome 1, planches 1 a 94 sont complets en 20 livraisons.

Le tome 2 comprendra les livraisons 21 a 40, avec les planches 95 a 200.

36 livraisons sont en vente. Prix de la livraison 8 75

LEURET ET GRATIOLET. Anatomie comparee du systeme nerveux considere dans ses rapports avec l'intelligence. 2 vols. in 8vo., et atlas de 32 planches in folio. Paris, 1839-1857.

Figures noires 12 00

Figures coloriees 24 00

LUDOVIC-HIRSCHFELD ET LEVEILLE. Nevrologie ou Description et iconographie du systeme nerveux et des organes des sens de l'homme, avec leur mode de preparations, par M. le docteur Ludovic-Hirschfeld, professeur d'Anatomie a l'ecole pra-

tique de la Faculté de Paris, et M. J. B. Leveillé, dessinateur. Paris, 1858. Ouvrage complet. 1 beau vol. 4to., composé de 400 pages de texte et de 92 planches 4to., dessinées d'apres nature, et lithographiées par M. Leveillé.
Prix: figures noires. Half bound 18 75
    figures coloriees 26 50

MANDL. Anatomie microscopique, par le docteur L. Mandl, professeur de microscopie. Paris, 1838-1848.—Cet ouvrage forme deux volumes in-folio.
Le tome I, comprenant l'HISTOLOGIE, et divise en deux series: *Tissus et organes.— Liquides organiques.* Il a ete publie en 26 livraisons, composees chacune de 5 feuilles de texte et 2 planches lithographiees in-folio 30 00
Le tome II., comprenant l'HISTOGENESE a ete publie en 20 livraisons . . 30 00

MANDL ET EHREMBERG (C. G.) Traite pratique du Microscope et de son emploi dans l'etude des corps organises, suivi de recherches sur l'organisation des animaux infusoires. In-8, avec 14 pl. Paris, 1830 . . . . 2 00

NOEGGERATH AND JACOBI. Contributions to Midwifery, and Diseases of Women and Children, with a Report on the Progress of Obstetrics, and Uterine and Infantile Pathology in 1858. 8vo, New York, 1858 . . . . . . 2 00

NELATON. Elemens de Pathologie Chirurgicale. 5 vols. 8vo. Paris, 1844 to 1859 . 9 25

NYSTEN. Dictionnaire des termes de medicine, de chirurgie, de pharmacie, des sciences accessoires et de l'art veterinaire, avec le synonomie Latine, Grecque, Allemande, Anglaise, Italienne, et Espagnole, suivi d'un vocabulaire de ces diverses langues. 1 tres fort volume de 1672 pages, avec 582 figures. Paris, 1858. ½ relie, maroquin . . 5 50

OTTO (J.) Manual of the Detection of Poisons by Medico-Chemical Analysis. By J. Otto, Professor of Chemistry in Brunswick, Germany. Edited with Notes by W. EIDERHORST. 12mo. With illustrations. New York, 1857 . . . . 1 75

OWEN. Odontography; or a Treatise on the Comparative Anatomy of the Teeth, their Physiological Relations, Mode of Development, and Microscopic Structure in Vertebrate Animals. By EICHARD OWEN, F.R.S., Corresponding Member of the Royal Academy of Sciences, Paris and Berlin; Reade's Lecturer in the University of Cambridge; Superintendent of the Natural History Department in the British Museum. In consequence of the small number remaining of the 8vo. edition of this work, the publisher has determined to reduce the 4to. edition, 2 vols, India paper, half Russia, published at £10 10s., to 35 00

PRICHARD. The Natural History of Man; comprising Inquiries into the Modifying Influences of Physical and Moral Agencies on the different Tribes of the Human Family. By JAMES COWLES PRICHARD, M.D., F.R.S., M.R.I.A., Corresponding Member of the National Institute, of the Royal Academy of Medicine, and of the Statistical Society, etc. 4th edition, revised and enlarged. By EDWIN NORRIS, of the Royal Asiatic Society, London. With 62 plates, colored, engraved on steel, and 100 engravings on wood. 2 vols. royal 8vo. elegantly bound in cloth. London, 1855 . . 10 00

———— Six Ethnographical Maps. Supplement to the Natural History of Man, and to the Researches into the Physical History of Mankind. Folio, colored, and one sheet of letter-press, in cloth boards. 2nd edition. London, 1850 . . . 6 00

QUEKETT (J.) A Practical Treatise on the Use of the Microscope, including the Different Methods of Preparing and Examining Animal, Vegetable, and Mineral Structures. 8vo. 3rd edition. London 5 00

———— Lectures on Histology, delivered at the Royal College of Surgeons of England. Vol. I. Elementary Tissues of Plants and Animals. Vol. II. On the Structure of the Skeletons of Plants and Animals. 2 vols. 8vo. Illustrated with 840 wood-cuts. Lond. 5 75

———— Descriptive and Illustrated Catalogue of the Histological Series contained in the Museum of the Royal College of Surgeons of England, prepared for the Microscope. Vol. I. Elementary Tissues of Vegetables and Animals. Vol. II. Structure of the Skeletons of Vertebrate Animals. 2 vols. 4to. Illustrated. London . . 17 50

RECORDS OF DAILY PRACTICE: a Scientific Visiting List for Physicians and Surgeons. This little book is not intended to supersede the use of a regular visiting list; its aim, as its title indicates, is to supply a medium for taking notes of the state of a patient, as soon after the visit as it is possible, and whilst the facts are still fresh in the memory. In hospital practice, we believe it will be found invaluable. Price, in cloth, 50 cents, with tucks 00 75

ROBIN (CH.) Du Microscope et des Injections dans leurs applications à l'anatomie et à la pathologie. 8vo. Paris, 1849 . . . . 1 75

———— Histoire Naturelle des vegetaux Parasites qui croissent sur l'homme et les animaux vivants. 1 vol. 8vo. et Atlas de 15 planches. Paris, 1853 . . . 4 00

SICHEL. Iconographie Ophthalmologique, ou Description et figures coloriees des maladies de l'organe de la vue, comprenant l'anatomie pathologique, la pathologie et la thérapeutique, medico-chirurgicales, par le docteur J. SICHEL, professeur d'ophthalmologie, medecin-oculiste des maisons d'education de la Legion d'honneur, etc., 1852-1859. Ouvrage complet, 2 vol. grand in 4to. dont 1 volume de 840 pages de texte, et 1 volume de 80 planches dessiinees d'apres nature, gravees et coloriees avec le plus grand soin, accompagnees d'un text descriptif 44 00
Le text se compose d'une exposition theorique et pratique de la science, dans laquelle viennent se grouper les observations cliniques, mises en concordance entre elles, et dont l'ensemble forme un *Traite clinique des maladies de l'organe de la vue,* commente et complete par une nombreuse serie de figures.

THE LONDON MEDICAL REVIEW (monthly) 8 60

VOGEL AND DAY. The Pathological Anatomy of the Human Body. By JULIUS VOGEL, M.D., Translated from the German, with Additions, by GEORGE E. DAY, M.D., Professor to the University of St. Andrew's. Illustrated with 100 plain and colored engravings. 8vo. London, 1847 . . . . 4 50